WONDER

When and Why the World Appears Radiant

Paul R. Fleischman

ALSO BY PAUL R. FLEISCHMAN

✦ The Healing Spirit: Explorations in Religion and Psychotherapy

✦ Spiritual Aspects of Psychiatric Practice

✦ Cultivating Inner Peace

✦ Karma and Chaos

✦ An Ancient Path

Poetry

✦ You Can Never Speak Up Too Often For the Love of All Things

✦ Masala Mala

ISBN: 978-1-937650-23-0
LCCN: 2013933247

Distributed by

SMALL
BATCH
BOOKS

493 SOUTH PLEASANT STREET
AMHERST, MASSACHUSETTS 01002
413.230.3943
SMALLBATCHBOOKS.COM

CONTENTS

For Susan

✦

Oh my love for the first time in my life,
My eyes are wide open…
John Lennon

✦

Where I lived was as far off as many a region
viewed nightly by astronomers…a forever new
and unprofaned part of the universe.
Henry David Thoreau

✦

Preface

I decided to write this book after asking myself two questions: What is the most valuable contribution that I can make to the fraught world? And, what do I want to spend my last years, months, days or hours thinking about, feeling, and doing? To both of these the answer was, wonder.

The topic that I find most urgent and interesting is the relationship between myself and the world, between the visions of cosmology, molecular biology, the evolution of life, and my mind. Some people search for their identity in ethnicity or gender. I wanted to locate myself as I am, in the context of M31, the "nearby" spiral galaxy pinwheeling fifteen quintillion miles away; and in the context of the short arm of chromosome 15, which silently formats me from within on the basis of its four billion-year-old chemical memory.

This exploration of wonder has four sources.

First, there is literature, poets like Whitman, Neruda, Keats, Dylan Thomas, and many others, who can evoke our wonder by describing theirs. We are not alone in our wonder, and have many friends and advisors, as well as some challenging mentors. Prose-poet Herman Melville, for example, beckons to us, mocks us, and reels out the story of his own courage, irreverence, wonder, and inconclusiveness. This book is written in reference to wonder-filled writing.

Second, there are the *discoveries* of science, which describe the world with the depth of genius compounded over centuries. Science is particularly revealing when it sweeps across disciplines to describe a whole. The movement of thought from our tiny interior cells out to cosmological origins and edges becomes numinous. Inside our cells are particles whose numbers and transformations echo galaxies and reveal something cosmological in the minute. The nature of life that molecular and cell biology can reveal to us (when these subjects are freed from memorizations and exams) is a glorious tapestry of our deep identity.

Third, there are the *ideas* of science, not its facts but its great concepts of cause, mutual influence, how the world is connected and held together. Our wonder at the world springs up from every bee and flower, the panorama of life, and also from its interconnections, which we now have a science to reveal as never before. We are embedded within a creative creation. We ourselves are new forms of emergence. Our home among the stars is lawful yet not predictable, impersonal but guided. Just as good citizens delight in benign laws, we are struck with wonder by the format of the cosmos.

The fourth source for this exploration of wonder is meditation on the arising and passing of the particles in my own body. This book partly derives from my lifelong, practiced awareness that every thing is an original, impermanent compound of disappearingly smaller parts. Everything is built and dissolves instantaneously in continuity. For the pervasiveness of this recognition, I am indebted to meditation as taught by the Buddha, but this book is neither about meditation nor Buddhism, nor am I a Buddhist, any more than the fact that I count on gravity to keep my feet on Earth makes me a "Newtonist." Meditation is cultivated awareness within oneself of change according to cause, and it has prepared me to witness the wonder of the working of the world. Yet this book is not limited to the legacy of meditation, and could not have been written without the scientists and writers whose work has formed the basis for the text to leap from internal experience into language and concepts.

To write about wonder, I have used sources from literature, science and meditation, but I have also been lucky. Wonder is more accessible to people who have been freed from superstition and coercion. The open-mindedness and partly-off-balance stance of wonder can feel intolerable to good but frightened people. Wonder emerges from a certain degree of confidence coupled to a kind of unknowing, and it is blackened out of existence by anxious entrenchment or conviction. I have been lucky to be granted intellectual opportunity in a relatively secure time.

By dwelling on wonder for so long, I hope I have added to the world and to myself a touch of liberation. Wonder is, among other things, a wand to dispel ignorance and appropriation. I remain inspired by the hope that truth will make us free.

Finally, it should be clear that I am not proposing wonder as vapid gaping. A wonder-drenched life is not denial of war, poverty or the stampede of collective human ignorance. The ability to feel wonder is a blessing received by someone whose ancestors and contemporaries struggled to throw off penury, ideology, and the pressure of the herd. I feel not only indebted and lucky, but grateful to the great phalanx of heroes, known and unknown, who pried open the door for me: writers, scientists, politicians, soldiers, immigrants, lawyers, protestors, contemplatives, and lovers.

If I contribute to a wider attunement to wonder, I hope this will also facilitate in its own way the reduction of violence directed at people and other lives. Who would destroy a breathing source of one's own wonder? I hope to bring into words a recognition that every white pine is a semaphore waving from the origin of time. Inside its tall trunk are atoms and laws much older than the Earth. Wonder and reverence are the royal marriage.

I owe a debt to Dr. Ted Slovin, who listened with such intensity, enthusiasm, and honesty, demanding clarity and evoking color. He heard me, helped me to hear myself, and provided many corrections and useful references. Chitra Amarasiriwardena, Manish Deopura, Rick Crutcher, Karen Donovan, and Susan Fleischman took the time to carefully read the manuscript and offer suggestions.

Just before I go to sleep and when I wake up at 5:00 AM, I try to catch a glimpse, through my well-positioned bedroom window, of stars which are unimaginably distant lights among which we sail without anchoring reference. An absurd and dismaying logic locates us within a universe of galaxies, black holes, and quasars, billions of trillions of miles wide.

Out of the immeasurable cosmic vaults of the universe, matter, energy, and information have converged to orchestrate us.

All of the massive evidence of the great twentieth century points towards nothing that resembles a person guiding the universe, and reveals equally forcefully that somewhere in its deep pulsations the world has directive. There are rules that act like barriers, currents, and channels that say "yes," "no," "maybe," "sometimes," "never," or "later" to the flow of events. There are signals by which everything to some degree has had limits set on what it must or cannot do. Something has touched and placed each one of the octillions of atoms that currently reside in me, and that have been guided there during a long cosmic dream. Order, loose patterns, and varied melodies have put in place the pieces out of whose coherence each one of us arises.

Weaving among the governors of the world like gravity, electromagnetism, or informatic constraint, among the stitches holding the seams of the universe, I believe, are threads of mind and wonder.

More than any idea, it is the irreducible complexity and incomprehensible presence of our long-woven awakening on our green Earth among cobalt galaxies, that impregnates us with wonder.

Introduction: From the Cell to the Galaxy

There is a large gap between the modern scientific worldview and the way we typically construct personal meaning. What does our life mean in a universe that is so big? When the Hubble Space Telescope, which orbits around Earth on a satellite, takes an ultra deep field picture that delves into an area with an opening the size of a straw, the penetrating photo reveals thousands of galaxies, each containing billions of stars, and yet there are uncountable equally straw-sized circumferences with equal galactic populations spanning the sky. How can we relate to such an ungraspable spatial context? Or, what does our existence imply on our planet that generated dinosaurs for more than one hundred million years, but then threw away all those ingenious and lugubrious inventions? Won't we also be thrown away? How can I relate myself to the scientific narrative in which I became immersed as a medical student, that I consist of trillions of tiny cells, each one of which is the product of billions of years of experimentation and molecular evolution? Am I too intricate to ever understand myself?

It is difficult to construct satisfying narratives that bridge our sense of ourselves to the realities revealed by science. Our failure to generate coherent new visions allows the human community to perseverate in archaic fables that have no truth value, and that are also filled with self-righteousness, divisiveness, and wishful thinking. The old fables are not only wrong in the way they describe time, space, and the evolution of levels of complexity; they also instigate false self-importance, which in turn triggers appropriation and aggression. When people cannot accept what they know, they proclaim what they wish to be true in a threatening tone of voice.

Then again, not everyone wants to be a scientist, and even science often presents temporary viewpoints that are soon discarded.

We want not only truth, but life-giving truth that can inspire and buttress us. Yet so much of what passes for "inspiration" is merely self-serving fantasy, hardly believable to those who desperately cling to it. Isn't there a way to find comfort, inspiration, and adventure in the context of our new, empirically-tempered world views?

Wonder is a form of intellectual innocence, coupled to a quest. To live with wonder, one must persevere in unknowing, re-encountering and participating. Wonder is not empty-mindedness, but an energetic engagement. We are born out of the world and contain within ourselves all of its processes which are activated in us as we meet them face to face in the course of our day, so that when we encounter it fully, the world awakens our reason, intuition, emotion, and wonder.

In moments of wonder we simultaneously contain a search for truth, an openness to reawakening, and a delight in what is. It is like our eyes. We see all day, yet we always want to see more. We want to see something new every day. Nothing lifts our spirits better than new scenes and fresh landscapes, but at the same time, nothing we see is entirely new, and we count on our sojourn in, say, Herefordshire, England, to reveal, along with its mossy churches and cockeyed hills, the same old gravity we know in America. Along with the new, we want enduring realities that help us feel at home in the world. We delight in both sacrosanct truths and fresh revelations.

When we dilate with wonder, we activate a unique form of complex knowing that integrates many kinds of awareness, thought, and feeling. The experience of wonder is itself a remarkable creation by the world within us, as I hope to show. We crave narratives confirmed by evidence, that are uplifting, and that bring our lives to a sense of fullness.

I hope to present science in an authentic and readable way that affirms the spiritual innuendos that remain suspended among the webs of statements and evidence. This is not a mere science book, but it is a consistent-with-science book.

We will steer between three kinds of closed minds and glide down the wide expanses between them. The first zone of closure is to imagine that the universe is run by a person, or by something like a person, or to imagine that the organization of the universe, from galaxies to cells, reflects a being something like your father, or something like the confabulating author of patently ignorant books. If you try to imagine yourself making galaxies of billions of stars spread over billions of trillions of miles, or, conversely, if you try to imagine yourself making a minute cell containing tens of thousands of complex, nearly perfect molecules, these imaginings will reveal how unlike we are to the forces that originated and organized the galaxy and the cell. Anthropomorphic creation fantasies diminish and distort the world.

The second way to close your mind is to conclude that the world is a machine, fixed by law, purposelessly clanking, making and unmaking matter and rag dolls like us. If you imagine yourself to consist of atoms and molecules organized by mere fate and chance, and you think of the octillions of atoms in your own body so exquisitely, skillfully interacting, you can quickly see what a "leap of faithlessness" the belief in "meaninglessness" is. An assertion that the universe is empty or void is no more scientific than a statement that the universe was made by a bearded man in the sky. Many of the most basic questions that we ask remain beyond science, such as: "What is mind?" "What is the origin of life on Earth?" "Of the trillions of stars we estimate to exist, how many have inhabitable planets, and inhabited by what?"

As ignorant as we are, why would we cleave to any conclusion? One of the main points of this book is that the density of interactions of law and history within the universe is formative and creative at a level that to us remains trans-calculable and transcendent. Science is not well served when it is used as a source of dogma or of prematurely foreclosed conclusions.

The third kind of closed mind is giving up, no longer trying to wake up within the world.

Wonder glides past three errors: ignorance (beliefs), cynicism (atheism), and laziness (agnosticism). Wonder arises with the freedom to worship and revere evidence and inconclusiveness.

We do not now possess either the intellectual capacity or adequate information to claim that we can speak for the universe. We cannot say we have reached a full and final understanding of it. But we also recognize the woven complexity and emergent luminosity within ourselves and our world. The complexity I am referring to means the galaxies and the cells, with their long history, their detailed precision, their self-generating cause-and-effect loops. The numinosity I am referring to is the green, sun-catching leaf, or our curious minds, or the astronomer, or the visionary poet. These presences within the world, these beings, constitute mirrors which have emerged from the universe and which look back at it with kinship, awe, dread, love and wonder. Through us the universe can become alive and gaze at itself with both alienation and recognition.

Therefore, our sense of wonder will lead us in the following direction: we will look at the world in a way that is free from the arbitrary divisions imposed by our routine sense of time, space, and scientific discipline. All the connections we are aware of will be permitted to emerge. At the heart of the evocation of wonder is that sweeping emotion, that stretching across, that connecting: origins to immediacies, light years to centimeters.

In Section I, voices of poets and scientists will evoke the sense of wonder. Their diverse perceptions make a "snap," like a window shade rolling up in the morning. While they share the sense of wonder, they do not share any common conclusion about the nature of reality or of the world. Section I is intended to open the windows and let in air and light. When we listen to our ancestors in the pursuit of wonder, we are asked to tolerate some degree of dismay and confusion, but we gain a bracing and energizing renewal, with freedom of thought and freshness of perception. Section I provides the necessary spring-cleaning. Although this Section weaves together writers' voices, it is not a précis or a review. It is the first step at the start of a journey, the curtain rising.

In Section II, I try to answer the question: What exactly is the sense of wonder? This is a psychological question about a state of mind, but my answer requires locating our psyche itself, in our bodies, on our planet, within the world. Our wonder is a localized form of a property of the world about which we wonder. Something about the world itself awakens in us when we feel the sense of wonder. Wonder is a type of complex awareness which participates in complex knowing and feeling.

Sections III, IV and V form an integrated whole, a journey through our best contemporary insights about ourselves. We, within whom the sense of wonder resides, must be studied, in order to understand how we respond to, mirror, and hold the world in wonder. These three sections are scientific, a kind of medical school for an accurate biology of wonder, because I hope to show how life itself is intrinsically connected to the sense of wonder. We are highly dense centers of atoms, energy, and information, compounded over long eons, following laws, opportunities, and vicissitudes. We temporarily cohere as people, due to the regime of the universe, which embraces both order and fresh happenstance. We are temporary expressions of the same creative processes that wove everything else, and that will continue to shred, renew, and re-create everything. We contain all the necessary ingredients, and the deep recipe. Wonder surges out of any study of our own molecules, cells, and processes, because of their unimaginable dimensions of time, numbers, and complexity. I hope that by the end of the empirical probing of our molecular nature that concludes Section V, wonder will no longer seem to be just one state of mind, but the most accurate and vibrant display of our experiences. It is within our molecule-made body that wonder bursts forth.

Wonder is the light by which the world is best revealed...though not completely! There are important reasons why wonder feels like our truest self and our most accurate monitor, and why we can't "finish it off" with finality.

Section VI, "A Constitution of Wonder," provides a reorientation. In the way that a political constitution provides a framework for society, a simple enough yet elastic enough rulebook, Section VI summarizes and tabulates the insights within our modern, lit-up world of wonder. It provides twenty viewpoints to keep wonder in focus. It summarizes the previous chapters in a manner that will make most sense to someone who has already absorbed their more global descriptions. Section VI anchors the previous chapters, but not too tightly. Wonder can in fact become our inseparable companion, but it can't be leashed or made to walk in little circles. Our traveling companions always include uncertainty and perpetual expectation.

Section VII attempts to evoke wonder from the immediate texture of human life. After traveling through light years, eons, molecular biology, and whales, it would be beneficial to bring our wonder down into the realm of our day-to-day psyche. Rightly perceived, the people we love and know best become our fountains of wonder. Paraphrasing the quote from novelist Thomas Wolfe that appears in Section I, Chapter Three, we can say that every person is a window on all time, and here is my person...

Every person consists of atoms that have been sorted and arranged. The energy for this task explodes out of other atoms as they are being melted inside of suns, and then it ripples down to us as sunlight. In the black space between the sun and us, the world undulates, and the waves of sunlight convey energy that can be used on Earth to *bond, communicate, create,* and *transform.* Energy has pulsed through the void, and bathed the Earth, and due to this glow the Earth has had the power to rearrange atoms in uncountable magnitudes, over eons, until the atomic world has been reshaped into whales and women, astronomers and novelists. Everything we see and touch consists of *matter rearranged* by *information and energy.*

Everything is in connections and bonds.

Life, mind, and love, our human nature, have been created in partnership with the rules that run the world. Our sense of wonder, which is a state of mind inside of a funny mammal, has within its grout the whole antecedent universe, its atoms, suns, electromagnetic waves, planets, and long-cooked constellations of cells and molecules that all are within us who feel the wonder. Our personal wonder is the local glow from a cosmic fire. Just as the Earth gets sunlight, we get in our mixture of chemicals the capacity for wonder. Within our thoughts and feelings, small components of the universe itself are being rearranged.

The mind that sees and understands the star is no less radiant than its object.

SECTION I

Voices of Wonder

Chapter One

On The Seashore of Endless Worlds

"By the toil of others we are led into the presence of things which have been brought from darkness into light."

Seneca (C.D.N. Costa, Tr.)

Many thoughtful people before us have been wonder-struck. What did they say? Why are some people so filled with wonder that they become spokespersons for it? We have around us, like elders seated at our campfire, these exemplars of the life of wonder.

The word, "wonder" is commonly associated with Rachel Carson, who wrote *A Sense of Wonder* while she was dying. At the end of her life, that had included the global controversy provoked by *Silent Spring*, her triumphant critique of pesticides, which made her the first effective world-stage eco-activist, Carson narrowed her final attention to one boy, and to her beloved Coast of Maine. She felt wonder radiating from nature, from the ocean, trees, moon, and moss, a state of mind that she could share best with a child. Although she felt wonder spontaneously emanating from the natural world, she also understood that wonder is a socialized perception, a communication between people and across generations. Wonder feels natural but is also partly cultural. Rachel Carson wrote, "A child's world is fresh and new and beautiful, full of wonder. It is our misfortune... that true instinct...is dimmed and even lost...If I had influence with the good fairy who is supposed to preside...I should ask that her gift to each child in the world be a sense of wonder so indestructible that it would last throughout life...." Her book brims over with pending death and her desire to transmit her spiritual legacy to one, and then to all, true disciples, about "...where the land slopes upward from the bay and the air becomes fragrant with spruce and balsam." She wanted her nephew, Roger, not so much to know, as to feel and become "receptive to what lies around you," through all his senses.

Carson was not unique in associating wonder with childhood. In *A Child's Christmas in Wales,* Dylan Thomas plunges his hand "into that wool-white bell-tongued ball of holiday memories" and out comes an evocation of wonder. Thomas' reverie threads his adult sensibility through his childhood impressions, creating a boyish yet sagacious nostalgia:

"Years and years ago, when I was a boy, when there were wolves in Wales, and birds the color of red-flannel petticoats whisked past the harp-shaped hills…before the motor car, before the wheel, before the duchess-faced horse, when we rode the daft and happy hills bareback, it snowed and it snowed…Our snow was not only shaken from white wash buckets down the sky, it came shawling out of the ground, and swam and drifted out of the arms and hands and bodies of the trees; snow grew overnight on the roofs of the houses like a pure and grandfather moss…"

Thomas intoxicates us with a colorful, cozy world. His wonder arises from an adult's longing for the comforting domestic dimensions of childhood: time stretching out, and senses flooding into each other. Nature becomes an active playmate; the hills themselves are daft and happy. The snow is alive, and everything feels timeless and immediate. Fear is for adults who are ignorant and wonder-challenged. It would seem easy to feel wonder when your memories are pearls of poetry.

The most famous exponent of wonder, the poster-boy of wonder, is Albert Einstein, whose brilliance and subsequent persecution removed the stigma of romanticism or naiveté, and who insisted upon wonder's enduring value for the soberest enterprise. The cerebral mathematical scientist insisted that he, too, was at the core, another child of wonder. Einstein wrote (as quoted in *Einstein* by Walter Isaacson) "We are in the position of a little child entering a huge library filled with books in many languages. The child knows someone must have written these books. It does not know how. It does not understand the languages in which they are written. The child dimly suspects a mysterious order in the arrangement of the books but doesn't know what it is. That, it seems to me, is the attitude of even the most intelligent human being…" Einstein linked wonder to a childhood of books, libraries, knowledge, and incredulity, a meeting ground between the freshness of ignorance and the wisdom of the sage, who recognizes the limits of his own gnosis.

Einstein embedded in the canon of human thought as precisely as $E=MC^2$, his sparkling credo called, "What I Believe," which he specifically wrote to both fulfill and thwart the many people who wanted him to become either a prophet or a guru. (There are several slightly varying translations from German into English). In this credo, the master tells us where he will and won't allow himself to be positioned:

"The most beautiful emotion we can experience is the mysterious. It is the fundamental emotion that stands at the cradle of all true art and science. He to whom this emotion is a stranger, who can no longer wonder and stand rapt in awe, is as good as dead, a snuffed-out candle. To sense that behind anything that can be experienced there is something that our minds cannot grasp, whose beauty and sublimity reaches us only indirectly: this is religiousness. In this sense, and in this sense only, I am a devoutly religious man."

Einstein was treading between two pitfalls. On the one hand, he wanted to avoid the mechanistic, atheistic, dead vision of the universe that is often associated with science. Einstein was emphatic that he was not an atheist. On the other hand, he did not want to be misunderstood as believing in any known religion, or any anthropomorphic God. Einstein elaborated his credo in the third person in an attempt to preserve its spiritual innuendos, while making sure he had abolished any affiliation with the narcissism and superstition that he felt plagued organized beliefs:

"Individual existence impresses him as a sort of prison, and he wants to experience the universe as a single significant whole...This kind of religious feeling knows no dogma and no God conceived in man's image, so that there can be no church...based on it...It is the function of art and science to awaken this feeling..." (From *Ideas and Opinions*).

Although wonder has a religious flavor, for Einstein its domain was in the arts and sciences.

While Dylan Thomas presented us with a charmed village soaked in nurturing perfection, Einstein descended into the cradle, the infancy of wonder, and discovered an impersonal experience of space-time that was both eerily congruent with human thought, yet fundamentally beyond human thought. The essence of his position was in the question: why do mathematical formulas seem to capture reality? Why is the world lawful; how did the laws come to be; why does the internal activity of the human mind—the creation of math—fit the universe's regularities?

As much as Einstein insisted on wonder, he re-positioned himself away from mere astonishment. He was not vague or dumbstruck. He did not fixate on "Wow." He took a stance: there is something beyond, to which he applied the words, "Spinoza's God," a lawgiver, though not a participant in our daily lives. The human mind—even Einstein's mind—has limits of understanding. We will always remain children in a research library, indirect knowers. The ultimate, the sublime, manifests in the order of the universe, which we can know only through its inexplicable resonance with mathematics. Scientific truth consists of "...the smallest possible number of mutually independent conceptual elements." Einstein's universe was ordered and elegant. There was one irreducible unity. He wrote in *Ideas and Opinions:* "...Whoever has undergone the intense experience of success in the domain of rational unification of the manifold, experiences profound reverence for the grandeur of reason incarnate in existence."

Einstein said, "incarnate"—not that reason is part of the world, not that reason is useful in understanding the world, but, for Einstein, reason was inside of existence. When existence came into being, reason was already in it. The universe was actually a manifestation of Reason. Einstein felt, (as quoted by Isaacson) "a rapturous amazement at the harmony of natural law, which reveals an intelligence of such superiority that, compared with it, all the systematic thinking and acting of human beings is utterly insignificant."

Despite the fact that he felt the universe is a product of reason and intelligence, Einstein also believed there is a limit to understanding beyond which no person can ever pass "...unattainable secrets of the harmony of the cosmos." The human apparatus, mind and math, simply cannot grasp the code embedded in existence. Wisdom's goal cannot be a thought, but must pass out of the realm of reason into a feeling: "...the cosmic religious feeling." (As quoted in Isaacson and *Ideas and Opinions*).

Einstein bequeathed to us an irony: the master of the rational, who believed that the universe itself is essentially rational, also believed that rationality has limits which only feelings can transcend.

Once again, we find Einstein walking a middle path between what he saw as two errors. One error was to lack faith in the universe's ultimate, transcendent derivation from reason. The other error was to assume that we now, or ever will, fully understand "the mind of God." Our deepest connection must remain not a thought, but a feeling.

If we try to summarize Einstein's wonder-struck mind, we find a person in resonant connection with childhood feelings, who treasured those feelings and gave them centrality. He recurrently sought escape from the prison of self, to fuse with the universe in a cosmic religious feeling. He justified, or amplified those feelings, by applying the powers of intellect, reason in its most rigorous form, mathematics, to reduce the welter of the world to unifying, simplifying formulas, which revealed that the universe is in its essence an incarnation of reason. He was humbled and uplifted by this emotional and intellectual communion with the superior intelligence within the world. He named this superior intelligence within creation, "God," but did not try to placate, cozy-up to, or influence this remote and incomprehensibly superior other. He wanted, as quoted in *Ideas and Opinions,* to "purify the religious impulse of the dross of its anthropomorphism." He recognized a limit to human thought. Like a mystic, he communed with the reason incarnate in existence, but like a worshipper, he believed there is an unbridgeable barrier between himself and the secret core of the universe's ways. Because he revered but could not truly comprehend, he was left rotating in an orbit of wonder.

Einstein's awestruck humility, and his heartfelt sense of human diminution, made him one of us, an any person, despite his millennial intellectual status. He attracted so much adoration because he was at once as puzzled as the rest of us, while he was at the same time the quintessential genius.

Between the cradle and the grave, Einstein developed and changed.

Ironically, Einstein's wonder in old age became caged behind a linear determinism, which he himself as a scientist brought into question, and which was subsequently toppled during the twentieth century, and replaced by probabilities, multi-determinisms, paradigm relativities, cultural constructivism and many other modes of scientific explanations, which we will explore through the thoughts of many other writers in this Section I. Einstein in his last decades became stuck in the position of an orthodox believer in a rigidly mathematical clockwork universe. Was it age, or his science, or the horror of the Holocaust that buried the wonder-struck child? Einstein's need to affirm a clear, simplifying, mathematizable order within creation, a rational God, made him a determinist who, ever logical, even denied his own free will. He entrenched himself in refutation of the creative variabilities, options, choices, chances, and elaborations that form the basis of modern science. His tragic blindness, of the otherwise supreme humanitarian and scientist, became codified in his famous quip, that God does not play dice. He demanded of the universe fixed relationships, free of chance, which does not seem to be the way things work, as we will explore in detail. After all, it was the God of the Hebrew Bible who affirmed free will, and it was the science of quantum physics, which Einstein founded and then rejected, which revealed probabilistic oscillation as the foundation of the mighty world. In *The Expanded Quotable Einstein* he wryly sighed: "To punish me for my contempt of authority, Fate has made me an authority myself."

Is it true that as we age, wonder fades? Rachel Carson worried that was too often true; Dylan Thomas killed himself by drink before he could find out, and Einstein seemed to fulfill this gray prophecy.

The English romantic poet, William Wordsworth, built a career on the mournful conviction that wonder is a trait of youth that is destined to fade. He wrote in his famous "Intimations of Immortality,"

> There was a time when meadow, grove, and stream,
> The earth, and every common sight,
>   To me did seem
> Appareled in celestial light,
> The glory and the freshness of a dream.
> It is not now as it hath been of yore; —
> Turn wheresoe'er I may,
>   By night or day,
> The things which I have seen I now can see no more...
>
> Shades of the prison-house begin to close
> Upon the growing Boy...

Other writers blame the loss of wonder upon cultural developments. In a widely quoted and re-quoted aphorism, from his bestseller, *The First Three Minutes,* Steven Weinberg, who shared the Nobel Prize for physics in 1979 for helping to prove a portion of Einstein's unfulfilled dream of unifying scientific laws into fewer and more sweeping laws, pinned the loss of wonder on the growth of information that accompanies maturing individuals and maturing civilizations: "The more the universe seems comprehensible, the more it also seems pointless." Weinberg's aphorism echoed another legendary atheistic anthem by French Nobel Laureate Jacques Monod, who wrote in his influential book, *Chance and Necessity,* "Man at last knows that he is alone in the unfeeling immensity of the universe, out of which he has emerged only by chance."

Although Steven Weinberg's famous comment has made him the hero of atheistic skeptics, and a straw-man for theists to attack, great minds do not easily get stuck on one-sentence conclusions. Weinberg has also less famously written about science and meaning in his essay, "A Designer Universe?": "So there seems to be an irreducible mystery that science will not eliminate." Is it possible that this irreducible mystery occasionally, or even frequently, thrusts Weinberg (who demonstrated the unity of the weak and electromagnetic forces), into a mood of wonder? Does great science derive from the inspiration of wonder, or create new arenas of wonder, or kill it?

Probably the most revered scientist who ever lived was Sir Isaac Newton. Adulation for him has never dimmed, and twentieth century giants, like Einstein and astronomer Edwin Hubble, literally lived their lives in his presence, as his portrait hung on both of their walls. Yet Newton's personality has a very negative reputation. Stephen Hawking wrote, in an appendage to his *Brief History of Time,* "Isaac Newton was not a pleasant man...with most of his late life spent embroiled in heated disputes." Newton is described elsewhere as cranky, isolative, and arrogant, a sort of Asberger's enfant terrible. Furthermore, his enduring discoveries have been used to form the constitution for fixed, deterministic, reductionist science, which portrays a clockwork, impersonal, inflexible world. Yet if Newtonian physics has been erected as the centerpiece of a mechanistic and dead universe, it is the product of his diminished interpreters, not of Newton himself, who wrote (as quoted by Carl Sagan in *Cosmos*):

"I do not know what I may appear to the world; but to myself I seem to have been only like a boy playing on the seashore, and diverting myself, in now and then finding a smoother pebble or a prettier shell than ordinary, while the great ocean of truth lay all undiscovered before me."

There is our child of wonder again, inside the mind of the discoverer of calculus and gravity. Whatever personality problems he may also have had, the emperor of science, Newton, like Rachel Carson, Dylan Thomas, and Albert Einstein, saw the world with beginner's mind. Newton was a boy draped in a cloak of wonder.

There is a striking parallel in phrasing between England's great scientist, and lines of India's great poet. Rabindranath Tagore described in *Gitanjali* how wonder is born in a child's mind at a specific location:

On the seashore of endless worlds children meet...
They build their houses with sand and they play with empty
    shells...

For scientist Newton, and poet Tagore, wonder was associated with the seashore, the location that has the longest view, and the meeting of three phases of matter: solid earth, liquid ocean, and immeasurable sky. The seashore served both Newton and Tagore as an image for child's play among horizonlessness and primary unmediated wonder.

As science progressed in the nineteenth and early twentieth centuries, there was an initial outcry that cold reason was killing wonder, but those days are long past, and in the shadow of Einstein, Newton, and many others, science has reversed vectors and become the source rather than the nemesis of wonder. The old days, when wonder was falsely associated with romanticized "spirituality" or "religion," have receded, and wonder has opened like a sparkler under the match of expanding information. Wonder is no longer associated with soothing mythology, but with grasping reality. As wonder has become more synonymous with knowledge, it has become more amplified, rather than diminished, with personal age. We will encounter a cadre of distinguished scientists who, after a lifetime of careful studies that culminated in a blaze of recognition, finally in retirement allowed themselves to depart from factual scientific journals, and to write their own books of wonder. The scientific world-view has also expanded poetic wonder, which could at last throw off the strictures of ideologies and expand into direct personal experience.

Chapter Two

The Whole Universe in Its Variety

The nineteenth century American poet, Walt Whitman, was a revolutionary, who embedded into poetry the content of new scientific discoveries. Into the radical oceanic rhythms of *Leaves of Grass*, he inserted scientific astronomy and cosmology rather than theology:

> I am the acme of things accomplished, and I am the encloser of
> things to be…
> Immense have been the preparations for me.
> Cycles ferried my cradle.
> For room for me stars kept aside in their own rings.
> All forces have been steadily employed to complete and delight me…

Nineteenth century science exploded the biblical sense of time and space. Here were dinosaur bones and the stratae of rocks, and telescopically expanded skies, and measurements that revealed the immensity of our galaxy. The idea of a few-thousand-year-old world evaporated. Walt Whitman and his late nineteenth century contemporaries were forced to wonder about the origin and relevance of their individual lives in the context of large time/space dimensions. Many people attacked science and tried to make its dimensional revolution go away. Other people felt diminished by the coordinates science seemed to reveal. Whitman embraced a zone of expanded purpose, a more elaborate pre-planning for him. The big platforms of the world gave him a more wonderful stage on which to declaim his poetry. His bumptious wonder included, but was not tied to, science:

> When I heard the learn'd astronomer
> When the proofs, the figures, were ranged in columns before me…
> …I wander'd off by myself,
> Into the mystical, moist night-air, and from time to time

Look'd up in perfect silence at the stars.

A mixture of scientific information and direct experience allowed Whitman to see himself within the "Kosmos," "vast trackless spaces," even plate tectonics, "the ancestral-continents grouped together," galactic evolution, "there are millions of suns left," with its attendant wonder about everything. Whitman's wonder was like a shooting star that plummeted from the "Kosmos" and fell glowing at his feet.

A child said, 'What is the grass? fetching it to me with full hands;
how could I answer the child? I do not know what it is any more
    than he.
There was a child went forth every day,
And the first object he look'd upon, that object he became…

Whitman fused with a delightful and indefinite world of scientifically provoked, open-minded wonder. (All quotes are from *Leaves of Grass*).

As the number and refinement of scientific explanations expanded, the number of missing explanations in the web of truth also multiplied. Science stopped promising to explain everything, and instead, opened up cascades of questions. Within causal thinking, there is an infinite regression, by which events are explained by antecedent causes, which were events, and which require antecedent causes. This process led Whitman to wonder. He saw each leaf of grass as resulting from a causal web of expanded and illimitable space, time and history. It takes a whole universe to craft every detail within it.

All truths wait in all things…

To me the converging objects of the universe perpetually flow…
I believe a leaf of grass is no less than the journey work of the
stars.

So Newton's scientific revolution entered poetry and amplified wonder, based on the scientific dictum that all laws act in all domains. (Later, science, as we will soon see, challenged Newton's dictum of cosmic uniformity, and created a vision of the world that contains domain-specific laws and phenomena). In Walt Whitman's vision, one blade of monocoecious grass is an inheritor of the laws of the universe, grown from the light of the stars. Everything that has ever happened is needed to explain anything that is now happening.

The message of science, as expressed by the poet of the scientific age, was that webs of universal laws, repeating over eons, touch down in each microcosm in each moment.

The world is a matrix of intricate causal interactions. A blade of grass and the starry sky share the same laws, the same creative processes, and the same unanswered questions.

But this was not an entirely new realization. Before the scientific cultures of Europe and America, similar insights had already animated traditional thought in India. That the microcosm reflected and necessitated the macrocosm was realized through meditation, intuition, vision, and verse.

Rabindranath Tagore, India's first Nobel Prize winner for literature, 1913, wrote in *Gitanjali:*

The time that my journey takes is long and the way of it long.
I came out on the chariot of the first gleam of light, and pursued my
     voyage through the wildernesses of worlds, leaving my track
     on many a star and planet.

Tagore experienced his personal life (which Einstein, who later was to meet with Tagore, called "the prison") as actually existing in a long continuum, that began with the origin of light, and which had traveled the long rays of stellar and cosmic evolution. Although Tagore had certainly read Whitman, his own imagery was more influenced by the scriptures of India, in which the unity that Einstein had imagined he had glimpsed, and which he sought in vain to mathematize, was a long-accepted assumption. Speaking to the universe as if it were a person, Tagore marveled that cosmic law also operated in his personal life:

Thou didst not turn in contempt from my childish play among
     the dust, and the steps that I heard in my playroom are the
     same that are echoing from star to star.

The Western educated Indian poet, Tagore, the child of wonder, felt the same about the dust at his feet and the vault of the stars. His world was a self-communicating entity, every part of which contained the essence of the whole.

Today, Tagore is often criticized for sounding quaint and naïve. Isn't the universe just too big and too old to have any meaning for us? Isn't our life, as Jacques Monod wrote, a quirk of fate and chance? Tagore did not think so. For him, the web of cause and effect lived not only around, but inside us. Just as a leaf of grass could reveal to Walt Whitman the universe's journeywork, Tagore felt the entire cosmos holographically cohered in one child and one woman:

> What was the power that made me open out into this vast
>   mystery like a bud in the forest at midnight?
> …the inscrutable without name and form took me in its arms
>   in the form of my own mother.

The poetic world that Tagore derived both from Indian traditions and from Western science personified and recognized cosmic forces as the animating spark inside of everyone. The universe was not understood through a telescope, but by seeing one's own life, even one's own emotions, in their cosmic continuity and context. Know yourself well, and know the creation. Before you took birth, the creation was primed to hug you into itself in the form of your mother. The universe gives birth to us, and our mothers contain and express this cosmic caring and preparation.

His sense of cosmic unity incorporated Tagore into an interwoven panorama and microcosm, a sky and a nest. As a child, Tagore had been taught Sanskrit and wonder by his father, and was introduced to the *Bhagavad Gita*, an Indian scripture which portrays the world as a divine unity. The *Bhagavad Gita* says (as translated by Juan Mascaro):

> And Arjuna saw in that form countless visions of wonder: eyes
>   from innumerable faces… numberless heavy weapons… the
>   Infinite Divinity was facing all sides, containing all marvels.

> If the light of a thousand suns suddenly arose in the sky, that
>   splendor might be compared to the radiance…of the whole
>   universe in its variety, standing in a vast unity…

> Trembling with awe and wonder, Arjuna bowed his head.

In the *Bhagavad Gita,* the variety of the universe is reduced to a unity through a sense of wonder. This poetic vision, which inspired Tagore, echoes Einstein's mathematical effort to rationally unify the manifold into the smallest number of conceptual elements. It has been a source of wonder to ancient poets and modern scientists to try to see our existence as a derivative of, or a vessel containing, components so ancient, distant, and unendingly recompounded, as the entire cosmos. There is an overarching, stunning entity, out of which variety is a temporary format.

The *Bhagavad Gita* expressed an emotional shift. It did not invoke childhood in association to wonder. Instead, the *Gita* invoked war and mass murder! Arjuna, the character who has the vision of wonder, is said, in the above translation, to be "trembling with awe and wonder." Here the poetically minded translator, Juan Mascaro, has tried to get to the essence behind the literal "hrstaroma," which in Sanskrit means "with hair standing on end!" Arjuna has his vision of cosmic unity while waiting on a battlefield for a war to begin. The war, described in the ancient Indian epic, the *Mahabharata* (of which the *Bhagavad Gita* is a section) culminates in the death of Arjuna as well as most of the warriors on both sides. The *Gita* connected awe and wonder to cosmic time and personal death. Wonder was associated with the obliterating magnitude of the world.

For us today, these famous slokas of the *Gita* are indelibly linked to the first atomic explosion which, as its hair-raising power unfurled, reminded physicist J. Robert Oppenheimer, the director of the Manhattan Project, of these specific Sanskrit verses from the *Gita*, which he had learned by heart. Is there a dark wonder, a black wonder, a hair-raising wonder, which is simultaneously revealing, amazing, clarifying and horrifying, a warrior's and not a child's wonder? Was Arjuna's reverential wonder in the presence of divine, all-devouring time, in which all lives of all eras rush like moths to enter a burning flame, in which flames devour all worlds, in any way connected to Dylan Thomas's wonder as he rode the daft and happy hills bareback? Can wonder endure the impact of destruction? Can we retain wonder in the post-Holocaust era of nuclear war? Was Einstein's wonder crushed by the Holocaust?

Hundreds of years before the *Gita* was written, about two thousand five hundred years ago, in India, provocateurs of wonder chided their listeners into visions of personal death, serial ages of death, Big Bang universes, and serial Big Bang universes. The Buddha is quoted in the *Bhayabherava Sutta*, (section 4, line 27, of the *Majjhima Nikaya*, as translated by Bhikkhu Bodhi):

"When my concentrated mind was thus purified, bright, unblemished, rid of imperfection, malleable, wieldy, steady, and attained to imperturbability, I directed it to knowledge of the recollection of past lives, one birth, two births, three births, ten births, twenty births, thirty births, a hundred births, a thousand births, a hundred thousand births, many eons of world-contraction, many eons of world-expansion, many eons of world-contraction and expansion..."

In his book, *The Argumentative Indian*, Indian born Nobel Prize winner, Amartya Sen, pointed out that the "Creation Hymn" of the *Rig Veda*, the most ancient Sanskrit Text, maybe three thousand five hundred years old, first gave a description of the origin of the universe, but then discouraged the listener (this was a pre-literate, oral text) from believing it! Sen wrote: "…these grand explorations of every possible religious belief coexist with deeply skeptical arguments…radical doubts." The Vedic text ends with a statement and a question: This is the way it happened…or is it? Wonder is an ancient habit, to be found cohabiting with belief and doubt. Ancient Indian thought has been described like this (in *India*, by A. Eraly, Y. Khan, G. Michell, and M. Saran):

"If there is a single window into the mysterious world of Indian arts, it is via a holistic vision of the world—where every being and action is viewed in its relation to the whole universe, past and present. The fascination lies, as with Indian religion and philosophy, in exploring the relationship of the part to the whole, the many to the one, and the finite to the infinite… Man acquires special significance not because he is first among nature, but because he has consciousness and the ability to transcend physicality…"

Interconnections, relationships, contexts, dissolve those explanations that are simple, partly true, or partly satisfying. Wonder is connected to childhood, science, information, doubts, darkness, time, death, and to the dissolution of arbitrary boundaries. Wonder is experienced from fusings and blendings. Our picture of reality can be expanded. When it becomes more encompassing, less compressed, less parochial, it may also become haloed with wonder.

Are we today any closer than Walt Whitman, Rabindranath Tagore, the authors of the *Bhagavad Gita*, to understanding ourselves and the universe in a fundamental, causal way? Can we account for our personal existence given the fact that such an ancient, far-flung backdrop exists? Doesn't the kernel of the world hold Einstein's "unattainable secrets," "irreducible mysteries," both hair-raising, and daft and happy venues? Is the twentieth century Nobel Prize level physics that led to Stephen Weinberg's "pointlessness" a rigorous explanation, or an expression of fatigue and pique? If we do not want to fall prey to unjustified and unnecessary blind belief or cynicism, can we stand anywhere but under the caravansary of wonder? Have we matured away from wonder, or has the astronomical cosmic perspective plunged us more deeply into it? Do we know that the entire universe is purposeful or purposeless? Do we really know enough to appoint ourselves judges of the whole thing?

Chapter Three

New Magic in a Dusty World

For those of us who are inheritors of the 20th century revolutions in physics and biology, our beliefs are haunted by inflating dimensions. The inscrutable without name and form does not provide most of us a hugging mother. The scientific picture of the universe keeps on growing bigger and older.

Biophysicist Werner Loewenstein wrote in his book, *The Touchstone of Life*, that the ordering forces of nature began with time. "The arrows hailing from out there are long—some have been on the fly for fifteen billion years. These are the information lines issuing from the initial information state in the universe." (Today the universe is generally thought to be 13.75 billion years old). The universe is huge, ancient, and impregnated with laws. What Einstein called "reason" is currently re-constructed as "information." The world emerges from an ordinance.

There is information in the universe. It directs matter and energy; it is the traffic cop that dictates direction and behavior of atomic nuclei and of galactic clusters. Science names it: "the law of gravity," or "the electromagnetic force." These laws limit and constrain, guide and direct all things. They are information because they tell the stuff of the world what it cannot do, and what it must do.

Are we to assume that when the universe began there were already bits of information telling its components which way to go? Did "plus" already attract "minus?" How did that initial information get into the origin of the universe? Where does the guiding information come from? Was everything pre-determined? Imagine the information it would have taken fourteen billion years ago to predicatively sculpt not only our sun, and our planet, but every street and ice-cream cone in Newark, New Jersey. There must surely be some free play, some general rules within which freedom or accidents can also happen. Alongside of information, there must be lack of information, free will, chance or unguided and unpredicted events.

While wonder rests on a wholistic apperception, such as in the *Gita*, or in Whitman or Tagore's poetry, or in Einstein's "rational unification of the manifold," it is most powerful when it sweeps *us* up into it. We wonder about "it," the "world," the "universe," but we are most in wonder about *ourselves*. Exactly how did the fifteen billion year old information arrows that Werner Loewenstein says began with the Big Bang, form me? Does the "initial information state of the universe" compound information, or localize it, as might be necessary to create a creature as complicated and specific as I feel myself to be? How did the simple science laws at the origin of time get compacted into me? Am I merely a program printed out by cosmic information, or am I partly independent from, free of law and constraint?

Lewis Thomas, the doctor-writer, contemplating the idea that every detail of the existing universe could not have been pre-figured in the initial information state, and that therefore chance must also shape our world, wrote in the *Lives of a Cell*:

"We are infinitely outnumbered by all (of our possible but non-existent) alternatives...statistically, the probability of any one of us being here is so small that you would think the mere fact of existing would keep us all in a contented dazzlement of surprise." Over billions of years, the twists of circumstances may have created us, but, by logic, must have precluded all the theoretically possible other beings and scenarios that do not exist. Alongside our world are invisible shadow-worlds that never came to be. Almost all of what could conceivably be, isn't. Whatever exists here today is a specific, guided, reduced sample of the vast possibilities that were never actualized and do not exist. How or why are we, personally, woven into this lucky selection?

Astronomer Carl Sagan wrote in *Cosmos*: "In every ejaculation there are hundreds of millions of sperm cells, only one of which can fertilize an egg and produce a member of the next generation of human beings. But which sperm succeeds in fertilizing an egg must depend upon the most minor and insignificant of factors...If even a little thing had gone differently 2,500 years ago, none of us would be here today. There would be billions of others living in our place."

The same improbability of existence, and its attendant "dazzlement of surprise," was captured by Thomas Wolfe in the opening passages of *Look Homeward Angel*, the novel that was once hailed as the fulfillment of "the Great American Novel." Wolfe wrote: "Destiny …is touched by that dark miracle of chance which makes new magic in a dusty world. Each of us is all the sums he has not counted: subtract us into nakedness and night again, and you shall see begin in Crete four thousand years ago the love that ended yesterday in Texas… Each moment is the fruit of forty thousand years…and every moment is a window on all time. This is a moment."

Why, "forty thousand years"? In the less-than-a-century between the prose-poetry of *Look Homeward Angel* and the informational biophysics of Werner Loewenstein, our metaphor for time expanded from 40,000 to 14,000,000,000, or 15,000,000,000. But the incomprehensible wonder of our own existence remains, particularly when measured against the possibility of having been entirely precluded by the play of long-woven circumstances. Each one of us at the very least is a product of one-in-a-hundred-million sperm cells across uncountable generations, a set of minute and rapidly receding odds.

On the one hand we seem to be the product of cosmic law and information, and on the other hand, Lewis Thomas, Carl Sagan, and Thomas Wolfe remind us that we are the latest appearance of long chains of chance and luck, without which we would never have been born.

The appearance of the original cosmic information state is incomprehensible, and the play of chance that led to our personal debut is equally incomprehensible. There seems to be "an irreducible mystery" to the original information state that is incarnate in existence, and also to our own most unlikely and improbable cameo appearance.

The scientifically-informed cosmic long view has magnified wonder. Why didn't we end up on one of the roulette numbers into which no ball fell? Now that we understand that the Earth itself is just a happenstance in the cooling down of a supernova, how much more wonderful is Lewis Thomas, Thomas Wolfe, or us? Whitman wrote:

Before I was born out of my mother generations guided me…
For it the nebula cohered to an orb…
We have thus far exhausted trillions of winters and summers…
Now on this spot I stand with my robust soul.

What amazes us is not only the length and complexity of history, of pre-history, of cosmic law and accident. Our bodies are compounds. We are piles of little things. "We understand," University of Chicago Professor Neil Shubin wrote in his book, *Your Inner Fish*, "we are simply a mosaic of bits and pieces found in virtually everything else on the planet." To the confusing interplay of information and chance, we now are observing a third factor: the organization of massive amounts of material into coherent and functioning units, such as us.

Most of the phosphorous in us—and phosphorous is one of our essential ingredients—is leached out of planetary rocks, which themselves congealed from stellar and planetary evolution. We have many tiny pieces of phosphorus in us that came out of the universe, congealed on Earth, became concentrated in us, and, beyond mere concentration, became remarkably *coordinated*. One phosphorus-containing molecule, adenosine triphosphate, known as "ATP," is manufactured ten million times per second in each of our approximately one-hundred trillion body cells, (See Chapter Thirty-Nine) for an approximate total of 10,000,000,000,000,000,000,000 phosphate units turned over each second in our bodies! "ATP" is not the only phosphorous-containing organic molecule. (See Chapter Sixty-Three) *The speed and coherence within our bodily chemistry seems incomprehensible.* Can I really be based on such a frenetic dance of rock-fragments? It does not seem possible to coordinate so many bits, at such speed, to make someone I experience as coherent and unified as "me."

We seem to be awakened by science into a trinity of wonder. How did law, reason, and information become part of the world? Where does it come from? Why are actual events, like myself, so chancy, rare, accidental, unplanned, resting on one single event out of millions of possibilities? How is it possible that my improbable body holds, contains, and expresses such a deep dimension of law and order which guides the atoms in such refined detail?

How can the mix of lawfulness and order in my body coexist with the universe of luck and chance that was just described by Lewis Thomas, Carl Sagan and Thomas Wolfe? If the universe is run on chance, then why is my body so well run? If my body reflects a scientifically ordered universe, then why is caprice so influential?

Chapter Four

A Confederation of Causality

Our worldview today must take into account large magnitudes of ordering forces, such as those that integrate the rapid biochemistry of my body, alongside the improbable odds that I exist at all.

Along with rules and randomness, science also reminds us to consider *patterns*, or *webs of rules*, *networks* of ordering forces that may provide internal order to things, like the veins inside the maple leaf. The initial information state may be less like a rulebook, and more like the etching in the leaf. The initial information state may be less like a sentence, more like a lattice, within which everything else is suspended. The universe is sequential in time, and it is also spatial. Information and law are not only serial but simultaneous and contiguous. We are influenced and patterned by the past and by the simultaneous.

"To the naïve realist the universe is a collection of objects," says physicist Paul Davies in *Superforce*. (See also Chapters Twelve and Sixty-Five). But he adds, "the electron encodes information about a comparatively vast structure in its neighborhood...(the universe) is an inseparable web of vibrating energy patterns." (In Chapters Forty-Eight and Forty-Nine electrons will be explained and discussed in more detail).

The piling up of little things in such a rapid dance as constitutes me, may be built from top-down information, such as the creation of stars, planets, phosphorous cycles, and human bodies; but the ordering forces inside me may also be partly bottom-up information, that is intrinsic to the inner patterns in things. Order may not only be imposed, but it may be pre-formed. According to Davies, who was awarded the Templeton Prize for scientific contributions to spiritual understanding: "An elementary particle is not an independently existing unanalyzable entity. It is, in essence, a set of relationships that reach outward to other things."

So the building blocks of the world, like electrons, may not be "blocks" at all. They may swell and shape all things from within, more like rivers and streams than like "blocks" or "balls." The order of the universe, and of my body, may not consist entirely of bossing around atoms and making them adhere to rules. The order may be partly coterminous within the subatomic particles, like electrons, of which atoms consist. Some degrees or types of order may be intrinsic to matter. Matter, like subatomic electrons, may emerge with certain associative rules built into them. We are not just piles built up piecemeal by order and chance, but we are *relational nets*, systems, webs. Given one piece, many other pieces may spontaneously arrange themselves nearby, a network of self-assembly. This idea that Paul Davies has borrowed from physics, rests on an intrinsic relational awareness that may exist inside all of matter.

Victor Weisskopf, the physicist who is said to have been the leader who made MIT a brand name, wrote in *Science*: "Specific electron wave patterns, held in confinement, shape the world we see..." Thomas Wolfe may have felt himself to have been a lucky outcome of 40,000 years of births, deaths, wars, and migrations. Vicktor Weisskopf may have experienced himself to have been a pattern preformed by waves of energy that follow intrinsic, micro-laws. On the one hand, scientific laws may regulate atoms and electrons, but on the other hand atoms and electrons may be born with certain degrees of lawfulness that is part of their existence. Electrons and atoms are not just bossed around by laws like gravity and electromagnetism. They may be small-scale bosses on their own, carrying their own charisma with them.

Instead of providing us with final and polished answers about the universe's machinery, science challenges us to wonder about law, chance, complexity, and the very nature of matter itself.

We may well be molded out of overlapping layers or domains of scientific laws: like the big cosmic forces of gravity, or the coincidental and chancy happenings of history, or the fluid patterns, which exert their force within the deepest, invisible layers of electrons and other sub-atomic structures. We may be not only compounds of *things* but compounds of *types of laws*.

Why would we want our explanations about the causal nature of our bodies, or of the world, to delete any one of these arenas? The force-fields which hue and shape us may in fact extend through galactic light-years, or may exert their sway only in locales more minute than electrons, or both.

The universe, and our bodies, may well be structured from outside forces, nurtured by inner waves, and rolled like gambler's dice. The world we now contemplate is not just a matter of law versus chance, but of many types and domains of constraints and freedoms. "Scientific law" is not an entity, but a collection of types. There are many laws, and there are many kinds of laws.

How has this confederation of causality come about? Why don't the domains of laws contradict, crash, ruin everything? What about Newton's assertion that the laws of the universe are the same everywhere, and that the motions of the planets can be explained by the same gravitational principles as the falling of an apple on Earth? Newton's pivotal role in the scientific revolution was his claim that the universe is run by a single set of uniform laws. Can his revolution be revolutionized?

Our sense of wonder is provoked more, not less, by the attempt to explain the world scientifically. We seem to constantly find explanations that clarify one thing, but which then only provoke more questions. The universe follows laws, regularities and information rules, such as those discovered by Newton and Einstein. But within the working of law, there is enormous room for chance, like Sagan's image of one sperm cell among millions actually fertilizing the egg that became us. And there are separate laws in the domain of tiny things, like electron wave patterns, which may or may not limit, or influence, the role of macro, Newtonian laws.

Even one more factor in the attempt to understand the causal universe must now be evoked: our explanations may be subject to the error of our historical era. New "truths" or "laws" may one day make our current knowledge seem quaint and wrong. As science progresses, our current theories may be replaced by others. But there is certainly some order in the world. Our perceptions of order may be partly misperceptions, or social constructions, but they are not entirely invented. We are not merely oozing blobs in a random soup of colliding electrons and atoms. Why not?

Chapter Five

An Inlet from a Universal Flood

To previous generations, there were either gods or laws to run the world. There was also fate, or luck, the roll of the dice. Today we inhabit a psychological worldview of multiple options. We have more information, more ideas about cause, and more questions. We have many more facts but feel ironically less convinced or smug. We are encircled by the awareness that whenever we find ourselves wondering there are always two topics: the objects of our wonder, and ourselves, the locus of the wonder.

In the upcoming chapters, we will dive into the problem: when someone feels a sense of wonder, what is its biological basis, and what has been said about this wondering mind? We will listen to voices of people who are both eminent, and who addressed the issue of the site where wonder occurs, the human mind. As always, wonder is a bridge builder and stream crosser. We will listen to physicists, biologists, Nazis, humanists, Catholics, Nobel Prize winners, skeptics, believers, and astronomers, to set us ablaze with wonder about the wondering mind and the world from which it springs. We will consider wonder as a feature of the world that is activated in response to features of the world.

We will start with a scientist's scientist, who asked the simplest and hardest question. His answers have had a deep resonance throughout scientific disciplines. He asked the question that ended Chapter Four: Why is the subatomic world free of chaos, collision and disarray? Why is the subatomic realm of tiny things so apparently obedient?

Wolfgang Pauli was a charismatic, iconoclastic physicist who associated with Einstein, Oppenheimer, and every other major player in the physics revolution of the mid-twentieth century, which brought us bendable space and time, nuclear weapons, laser beams, and a more deeply understood and incomprehensible universe. His great insight, "the Pauli exclusion principle," helps to explain why, given the fact that gravity attracts everything to everything else, matter does not clump up into one dense and unmalleable ball. Electrons, of course, are tiny, even when compared to one atom, so they are, relatively speaking, far from each other. They also repel each other since they all carry a negative electromagnetic charge. Still, why are there never collisions or clumps? By studying the tiny world of electrons, Pauli located a counter player to the great gray power of gravity. Pauli's discovery also provoked a new depth to questions about the origins of order in the universe. Pauli realized something cosmic in the minute.

Pauli observed that electrons behave as if they were unique individuals owning their own spaces. They also act as if they are aware of what is going on around them. Because electrons seem to know what "orbits" they themselves occupy, and what electromagnetic signatures, or "spins," they carry, as well as what orbit and signature any other electron has, each electron inhabits an exclusive space, into which others do not trespass. This apparent spatial awareness holds the tiny world apart from itself, and prevents gravity-driven collapse. Electrons *exclude* each other. Electrons do not gravitationally attract each other into a dense, hard blob. Nor do they just electromagnetically repel each other to speed away into reverse direction. The Pauli exclusion principle accounts for the absence of collisional chaos in the whirling electron universe within all matter. (Electron exclusion may not extend to all other particle types, such as positrons.) Most atoms of which we consist, like carbon, oxygen, phosphorous, each contain many electrons, yet these siblings within one tiny atom never crash, and never drive each other away. In Chapter Four we heard Paul Davies describe elementary particles (like electrons) as "a set of relationships," and Victor Weisskopf describe electrons as "wave patterns that shape the world." Both of these descriptions derive from Pauli's principle.

Many scientists consider the Pauli exclusion principle to be one of our most profound insights into the fundamental nature of the Universe. Wolfgang Pauli was nominated for the Nobel Prize by none other than Albert Einstein, and he attained a status among his colleagues that some writers have likened to the "Supreme Court Justice of nuclear physics." The intimidating Pauli was alleged to have yelled at his poor PhD candidates: "Every time I say something, contradict me with detailed arguments!"

Specifically how does the Pauli exclusion principle help us to understand Lewis Thomas's dazzlement of surprise at his own mere existence, or the whirling dance of ATP and other chemicals in my cells? How does it help us understand the play of law, order, and chance? Pauli's abstract insight from physics has been pulled down into the arena of ourselves, our bodies and minds, by a great biologist.

One of the revered theorists of life sciences is Harold Morowitz. (See also Chapter Sixty-Six). Morowitz sees a very important role for the Pauli exclusion principle. He directs our attention to our minds. Morowitz believes that the human mind is not in fact quintessentially human. He describes "mind" as a property of the universe itself, which is held, or mirrored, in the human being, just as the other natural laws and forces keep our blood flowing or our ATP synthesis on track. Our eyes see light, but do not create light. Light is an antecedent property of the universe to which our eyes are only sensitive and responsive. Analogously, is it possible that our minds "know" but "knowing " is an antecedent property of the universe?

Morowitz sees the origin of noesis, knowing, prefigured in the Pauli exclusion principle, which reveals that even inside an atom there is self-awareness, and sensitivity to the "other."

According to Morowitz's interpretation of the Pauli exclusion principle, every electron has the equivalent of its own "mind." Awareness may well be a property of subatomic matter. Uniqueness and information seem intrinsic to the elementary particles of the physical universe. Electrons stay apart from each other, and keep tabs on each other. Because the state of one electron excludes any other electron from occupying that state, even the tiniest matter responds to relationships. Morowitz concluded: "(They) show in togetherness laws of behavior different from the laws that govern them in isolation." Matter may not be "dead" or "stuff," but at its essence, it may be aware of its environment. Awareness may not be a unique property of much-vaunted humanity. Awareness may be intrinsic to the universe, and we may merely harness and utilize it. Awareness may be a fundamental property rather than our unique signature possession.

My sense of wonder may be an inlet from a universal flood of awareness at cosmic high tide. My wonder derives from neural circuitry in my brain that can tap into a frequency of the world that wonders. Wonder may not be within me, but may wash up into me.

Even if awareness is a property of matter, we would still construe Dr. Morowitz's mind as somehow different from, say, the awareness in the electrons in the atoms inside a golf ball. To explain this difference,

Morowitz turned to his lifelong study of metabolism. As a biologist, Morowitz studied the universal features of metabolism. He was eminent in the army of scientists who, during the course of the twentieth century, have proven that every life form shares common biochemical processes. He wrote (in his book, *The Emergence of Everything*): "Here we sense the full impact of unity; the single *Chart of Intermediary Metabolism* applies with equal validity to all the millions of species...the core set of biochemical reactions of any organism, from a bacterium to a blue whale, is found on this four page chart...mind is not something that suddenly appears, it is a maturing characteristic of an aging universe." (See also Chapter Twelve).

In other words, life is the slow maturation of awareness during the evolution of all beings. Evolution not only diversifies species, but has also a core thread. Along that core thread is the expansion of awareness. Mind is a product of the shared and uniform core of life, metabolism. As metabolism evolved and matured from bacteria to whales and men, it became increasingly adept at capturing and utilizing "mind" that pre-exists in electrons and in all things.

"Mind" therefore derives from two domains: matter, electrons; and life, cells, metabolism. Our minds may be the products of metabolism maturing and sifting the cosmic mind, a process that has been slowly perfected by evolutionary adaptation during eons.

All of life is built on the same biochemical reactions, which, as they evolve, retain their core, but which also gradually mature in their ability to utilize the universe with greater capacity, sensitivity and skill. It may be that living minds, such as ours, do not create information, connections, or ideas. Mind may be a biological, evolutionary capacity to crystallize, or download, properties of the universe which are already present in the universe, in the exclusion principle and maybe elsewhere. All matter may contain the basis of "mind," and the development of life may represent the purification and amplification of that property. Mind may be less a creation than a distillate.

But then again, there is our old friend and nemesis, chance. Were "mind" and "wonder" destined to appear in a maturing universe, or were they stumbled upon by mere luck?

In contrast to what Morowitz thinks, "wonder" may well be a mental state that was selected by evolution under special conditions that were themselves *coincidental*. Wonder within my mind may be a chance outcome of evolution, like a frog. The less wonderstruck bacteria, who greatly outnumber and (in collective) outweigh me, did not require a sense of wonder in order to thrive, proliferate, adapt and endure for billions of years. Wonder is not a necessary outcome of adaptive evolution, nor is it necessarily valuable to survival.

But then again, it is difficult for a human to look back over the long development of life, without inserting a self-congratulatory sense of destiny. Wonder is a product of a complex mind that has been a long time coming, and here we are, at the top of the historical stairs. Isn't wonder the thoughtful quintessence of biological evolution? Isn't my sense of wonder a triumph of the evolutionary tree? Was it "the goal," or at least "destined?"

In Chapter Eight we will consider the question of how common or rare life is likely to be in the context of the whole universe. If awareness is intrinsic to matter, and if metabolism is destined to amplify awareness, and *if* life is common, then "mind" would certainly seem to drop down from cosmic destiny, like a pear from a tree. Wonder may be the universe seasoning like a fruit, "a maturing attribute" of an aging cosmos. The dangling universe has ripened into sweet wonder.

Why do we refer to the "human mind" at all? What makes us think that we "contain" our minds? Our eyes see light, but we do not speak about "human light" or try to puzzle out how "light evolved through apes to us." Light is a phenomenon we capture, record, and utilize, but it pre-exists independently from us. Is it possible that "mind" is similar? Do *we* think? Or do we observe thoughts, which are already "there"? Almost all creative scientists and artists feel their best ideas appear, come to them, arrive from elsewhere. In our best moments, are we speaking truth, or are we listening to it and merely translating it into our current tongue?

A ball on the top of a hill has potential energy even when it is sitting still. If the ball rolls down the hill the potential energy is activated, and the mass of the falling ball can act in the world to knock or crush other things. Could there be an analogous "potential mind"? It exists, but does not act upon things until it "falls down" into biological amplifiers, like brains. When our brains (derived from the evolution of metabolizing cells) receive mind descending from its pre-biotic, noetic potential, we think and can feel wonder. This is a rhapsodic idea, but far from a proven one. There are other theories about the mind that feels wonder.

There is the "quantum, a-causal, multiverse" viewpoint about the meaning and origin of the sense of wonder. Given the possibility that every possible combination of every possible atom can be generated in the context of infinite universes in infinite space over infinite time, maybe the wonder in my mind is *both* rare and inevitable. (See Chapter Thirteen). How many monkeys at how many typewriters are needed to randomly type out "Hamlet," and how many universes does it take before molecular neurology coheres into a state of wonder? While this playful "multiverse" model may descend into the fatuous, the Buddha himself envisioned cycles of worlds, eons long, pouring their diamond residues of Karma into the next Big Bang's information state.

I may wonder partly as a product of very long sequences of cause and effect that are quite specific to my personal karma. I may have personally evolved over many lifetimes in many universes to wonder. You, reader, may have a karmically crafted sense of wonder subtly different from mine, something to keep track of as you read this book chapter after chapter. Wonder may be a gift to each of us from the deeds of our personal past, idiosyncratically crafted across the histories of universes.

Maybe wonder is a newly emergent property of an old, old universe. Matter, energy, molecules, and neurons have cohered into organizational patterns that for the first time enable things of the world to wake up and ask. Like a man or woman, the maturing universe may become self-reflective. Wonder may be a beautiful, rare, biological maturation on one single planet at one point in time, the way that Thoreau's book, *Walden*, is a one-shot, one time event in the history of the universe. Maybe our human wonder is a diamond, a rare and sparkling exception to the dark, cold, dumb emptiness of the universe. Maybe we are stunningly, scarily unique.

Or our sense of wonder may be one misdirected evolutionary imperative, a blithe and romantic ricochet away from utility into daydreams: wonder may be a blunder. We humans, after all, are stumbling around our planet like dangerous and misguided apes. If Herman Melville could speculate in *Moby Dick* that most of religion is a product of indigestion, possibly the sense of wonder in a human may be little more than cerebral dysphoria. Wonder may just be a mental epiphenomenon of dizziness. Sometimes it feels that way. Wonder may be a biological misstep, like a platypus. Maybe wonder misleads us into speculations, philosophy, and religion that are poorly informed, deluded, dead-ends.

What is this mind of ours that both seems to *know* so intrinsically, and yet does *not know*, blocked by "unattainable secrets?" Our mind seems simultaneously part of and alienated from the wonder it beholds. Our minds are products of the universe; yet, Einstein said, we will always remain ignorant of the world's fundamental secrets. When we feel wonder, is that our unique human prerogative, or a property shared by Wolfgang Pauli's electrons and Harold Morowitz's blue whales? As we try to understand the world and the wonder it creates in us, we come back to the nature of our own wondering mind, and wonder about it.

Is our mind which wonders actually a property of a wondering universe? Could we experience wonder at all if it weren't a possibility within the antecedent potentials of the world?

The wondering mind at times observes itself in wonder. It sees itself seeing…is it, itself, embedded within the universe, or does it stand apart and touch into the universe with independence and detached perspective? Can the mind ever observe itself in such a way that it can confidently resolve such questions about itself? Is there ever "knowing" that is not distorted by the location and "mind" of the knower? Or is that knower utilizing, downloading, a cosmically embedded universal mind that has unimpeded access to the secrets of the essence and the core? If "mind" is connected to intrinsic properties of electrons and the micro-core of the universe, how did that "mind" get into the universe in the first place? How did the electrons become aware of each other's orbits and spins? Who or what, before Wolfgang Pauli, taught electrons their code of polite exclusion?

We are generally taught that our mind is in our brain, but now we are wondering whether the brain's role is to localize, focalize, capture, magnify, or intensify qualities of mind which preexist it. Mind may be a feature of the world which the brain reveals.

Chapter Six

Awareness and Creativity

One reason that we are left hanging in unanswered questions about something so intimate to us as our own mind is that "mind" is hard to define. We feel wonder when we look at the night sky, or when we contemplate the astounding number of biochemical reactions within us every second, yet we cannot clearly "feel" or "intuit" the mind within which that sense of wonder occurs. The wonder feels like "me" observing "it." The sense of wonder feels as if it were us, our center, our truest self, and the mind in which it arises is as difficult to observe as our own eyes.

What we mean when we say, "mind," is inexact. Do we mean awareness? In order for Pauli exclusion to occur, it does indeed seem that two electrons in nearby atomic orbits must be aware of each other. But maybe "mind" means *choice*? It seems a "stretch" to attribute choice to electrons. It is even difficult to attribute choice to humans, and for centuries determinists like Einstein have denied that humans really do have choice. Einstein and many others believed that all human actions are purely products of laws. It seems ironic that one of the most original men felt his thoughts were simply products stamped out by the cause-and-effect sequence of the universe.

It is thrilling and "mind-expanding" to consider the option that the universe itself has a mind, and that we do not contain our own mind, but tap into a much bigger one, which we download, or listen to. But there is also something unsatisfactory in this attribution of mind to electrons, matter, and bacteria, for we experience ourselves not merely as conscious, but as thoughtful, creative, *originators* of ideas and things.

A more restrictive, human-centered sense of mind might revolve around the trait of "hypothetical representation," the ability to imagine, envision, or think about something that is not actually present. Also,

human-centered mind might mean the ability to *re-order* these hypothetical representations in space and time, to construct theoretical variations in
patterns and consequences that do not exist, except in psychological representation. These two steps, representation and reordering, could form
the basis for goals, choices, planning, and creating. In this sense, mind is a
reorganizing phantom universe, a virtual puppet show of intangibles that
can be dreamed and shuffled only in our stunning human brains.

When we think of representation and reordering as the essence of
mind, we do not feel satisfied that the electrons in the atoms in the golf ball
have a "mind." Pauli showed us that electrons have "exclusion," Morowitz
re-interpreted "exclusion" as "awareness," and Davies called that awareness "a set of relationships." But now we are considering a very different
property of mind: its ability to create order. Mind may be thought of as not
only awareness, but creativity.

We experience our minds as capable of selecting, rejecting and reordering mental representation, and then *actualizing* the reordered images.
We dream, wonder, hope, fear and then act and try to rebuild the world
accordingly. Mind seems to exist at a boundary, where the things of the
world are pulled into a new order. Mind could be thought of as a physical-
chemical-biological boundary interaction where existence can incorporate
or dismiss alternative arrangements of the atoms and molecules that lie
inside of the things that get manipulated by actions. In this view, mind
seems "independent," "separate," an autonomous locus of local creation.
We humans often experience ourselves as little gods, free and self-directed,
able to remake the world. From within virtual, representational, mind-
space, we reorder and create worlds, which we then use as blueprints to
build.

Are our minds derived from "the original information state" of the
lawfully guided world? Is our sense of wonder, then, part of that original
information state? Surely, the Pauli exclusion principle must be part of
"the original information state of the universe." Or, on the contrary, is our
mind a free and separate center of local law and order, a creator in its own
right? To Einstein, this question was the key to science. He wanted to see
everything as a product of "the mind of God," the original information
state. His own sense of wonder was so radiant and adoring because he
must also have felt that his wonder about the mind of God was determined, caused by, the mind of God itself.

Yet our human mind also seems to draw up and expand order, to bring new alignments of molecules together in the houses, cars, and paintings that we make. We do not *feel* "determined." The vast, sprawling, cultural apparatus we have created appears to us to be "human-made," a product of choices and creations that cannot possibly have been pre-coded fourteen billion years ago, fourteen billion light-years away.

Mind now seems like an interactive space where the fuzzy, chaotic world of atoms and energy are given direction from *independent epicenters of selection*, rearrangement and freedom: us! Our sense of wonder may well be a unique, free, creative human invention, an undetermined mental state, a product of the same independent republic as the rest of our consciousness and choices. We may not be fruit in the orchard of the universe, but orchardists planting and tending our own crop

We now have two points of view. In one option, we experience our minds as downloaded from a mind somewhere in a cosmic "information state," "a mind of God." This type of mind is destined and determined, crystallized from pervasive, hidden processes. Einstein's mind felt wonder with the wonder-locus of the world. But we also can consider our minds as blazing centers of fresh invention. Our sense of wonder stands apart from and receives the impact of the world like a comet. The wide-eyed child in Dylan Thomas' Wales rides the daft and happy hills in fresh freedom of uniquely human apperception.

The roots of a plant finger their way into the soil and select nutrients, like phosphorous for their ATP. From among the atoms and molecules present around them, the rootlets choose, absorb, and concentrate. Then, out of tiny impersonal particles, the plant builds, re-builds, and maintains a plant. Does a plant create itself? Does a plant have more or less a mind than Morowitz's elections, or than a human artist or city planner? Analogously is our planet determined, or free? Selection and re-ordering are by no means unique to the human brain, although representation may well be.

The human brain carries out what the world of matter and plants can also do: select, arrange, order, and preserve. We drink order out of chaos, the way a flower drinks minerals from the soil and rearranges them into azure petals. However, we can do this both with our bodies and our minds. Our minds carry forward properties of life into new zones. Within our mind, selection and creation feel as if they occur in an abstract, independent creative space.

Our source of wonder is the awareness of the intricacy of the world. A heaven blazing with billions of stars, a body animated by immeasurable, coordinated high-speed biochemical interactions, make us aware of how many parts are relating to each other in a grand whole. Whether we use modern scientific knowledge or not, this means that we often are struck with wonder by an implicit question: how did so much information (arrangement, relationship, coordination) get into this picture? The mind that apprehends this information is similarly complex, ordered, and provocative of questions. So we also ask, how did so much information (arrangement, relationship, coordination) get into my mind so that it can feel wonder?

If it is true, as Pauli, Morowitz, Davies and Weisskopf say, that electrons are intrinsically aware, relational densities of information whose waveforms shape the world from within, where did the electron get this informatic orientation? If the "Big Bang" sourced gravity and electrons (sooner or later from its originating moment) where did the information within it, within electrons, within our minds come from?

Is the Big Bang itself a reliable truth, or just another scientific paradigm, destined to be replaced by another constructed, culturally relative narrative? The orthodoxy of the Big Bang is already under challenge, as we will discuss later in Chapter Twelve. The apparent order of the world, and of me, may derive from information coded in the Big Bang, or may be a consequence of wave patterns of electrons within all gross matter, or may be chance and luck, or may be a mix of my wishful perceptions and parochial distortions, which misconstrue degrees and origins of order. I may think there is more order than there actually is. Maybe I see order that I wish were there, but is it just my projection onto a fluid, fluxing chaos? Just how ordered is the order?

When we give free reign to wonder, we end up wondering about the mind in which wonder occurs, and the universe in which the mind occurs. The world becomes increasingly remarkable, unaccounted for, and counter-intuitive. There are brilliant and socially adjusted electrons orchestrating a world of profligate and millions-fold unlucky feckless sperm. It simply does not make intuitive sense that on the one hand the elementary particles of the world, and my mind, seem so smart and well run, while at the same time my entire existence is a highly improbable outcome of a series of giant marathons of randomly selected reproductive cells who won unlikely lotteries generation after generation.

Chapter Seven

The Veil

Many scientists now assume along with Einstein that human knowledge is intrinsically limited, and that final, absolute truths of the type we have just been asking can never be known, echoing Einstein's phrase, "unattainable secrets... a limit beyond which no person can pass...." Today this view is sometimes based on the uncertainty principle of Austrian Nobel Prize winner Werner Heisenberg, who proved that we cannot simultaneously know an electron's speed and position. If the tiny electron is measured for speed, it cannot be pinned down as to its place. If we measure precisely where it is standing, we can't also measure how fast it is moving. Having some information precludes us from possessing certain other information. The speed and the position of an electron are pieces of information that cannot simultaneously exist within the universe. It is not that some information is missing or difficult to obtain, but that our own position in the universe absolutely precludes our ever knowing certain things. Our stance intrinsically limits and biases us. Whenever we measure or observe something, our investigation alters or biases the conclusion. There is no objective fact, or neutral stance. Everything we can ever know contains some contamination from our location, our culture, our pre-conceptions, our actions, our historical moment.

Our inability to accurately and fully describe an electron now, in the moment, also means that we cannot precisely predict its behavior in the future. Heisenberg's uncertainty principle forms a cornerstone of quantum physics that is, paradoxically, almost unanimously accepted as "certain."

With similar impact, mathematical logician Kurt Gödel proved the inevitable incompleteness of all mathematical systems. (When Einstein was old, stuck in his futile attempts to find a grand, unifying equation for the "mind of God," and deemed irrelevant by other scientists, he said he went to his office every day at Princeton's Institute for Advanced Study, just to be able to walk home with Kurt Gödel). Heisenberg's uncertainty principle and Gödel's incompleteness theorem seem to impose (in the words of Stephen Hawking) "fundamental limitation on our ability to understand and predict." Because right now some facts are absolutely unavailable to us, we can never create predictions which extrapolate from the present into the future with full accuracy.

Heisenberg's uncertainty principle and Gödel's incompleteness theorem seem to be master discoveries that throw into question the project of knowledge. Does our sense of wonder boil down to an ignorant haze?

Of course, it is important to keep in mind that scientific laws and discoveries are domain specific. Giraffes account for grazing in Kenya but they do not account for hoof beats in Nebraska. Heisenberg was not claiming that Newton's law of gravity cannot predict whether a basketball will fall down, or will, upon passive release, spontaneously fly up into the basket for two points. Heisenberg's laws involved the quantum domain only, as formulated originally by Max Planck, which has specific dimensional limits, related to the size of Planck's constant, the arena of the extremely minute, mathematically precise "quantum." Heisenberg did not claim that everything is uncertain, or that nothing can be known. He was not implying that all reasonable statements about degrees of informational certainty had now been invalidated. He was not a "cultural constructionist" who claimed that all knowledge is a temporary local invention. He did not intend to imply that all knowledge is merely a culturally constructed narrative, and that if a person who is not a white, Anglo-Saxon male (as Newton was) holds a basketball and lets it go, it might fly up and smack her in the face.

Heisenberg *was*, however, a Nazi! He had joined the youth movement as a young man, to sing and hike mountains with the sons of the Fatherland. He was also very young when he established his reputation as one of the greatest mathematical scientists of all time, and the formulator of a postulate that has revolutionized the human condition. It is unlikely that Heisenberg became a Nazi to vent hate and anti-Semitism, since his mentor and closest collaborator was Neils Bohr, a Danish Jew. But Heisenberg did not flee from the Nazi war effort, as did so many of his colleagues, and he accepted leadership of the German nuclear weapons program. Due to Heisenberg's reputation for break-through brilliance, Einstein, who had previously considered preposterous the idea that his formula for the conversion of matter into energy could be used to create explosive weapons, now took the threat seriously, and personally informed Franklin Roosevelt, who immediately organized the Manhattan Project under the guidance of Oppenheimer. America, of course, won the race for the bomb, largely due to the help of European Jewish scientists who the Nazis had ironically alienated and catapulted into exile.

In 1941, when the outcome of World War II was still years in the future, Heisenberg made a visit, in official German uniform, to Bohr in Copenhagen, during which he alluded to his work on German nuclear weapons. Mutual misunderstanding between the two old friends occurred and subsequent historical controversy has further obscured what actually was said during this meeting. There remains speculation that Heisenberg implied that he was planning to undermine the development of a German nuclear bomb. This formed the plot tension in Michael Frayn's play, *Copenhagen*. Historical documents do not prove that Heisenberg actually implied this. Bohr's impression, based upon his private papers released by his family only in 2002, was that Heisenberg was happy to pursue the development of nuclear weapons.

Is it merely ironic, or a profound excavation of deep truth, that the founder of the uncertainty principle has also bequeathed to us so many uncertainties about his motives, his integrity, and the ultimate cause of the outcome of World War II? Wouldn't our world after 1945 be immeasurably different if the brilliant and stealthy Heisenberg had successfully led an effort to create a nuclear weapon, or if he actually had from the start undermined a bomb project which otherwise could have progressed further and faster? Have we plummeted from quantum domains to the cosmic dark wonder evoked by the mass destruction described in the *Bhagavad Gita*? (Chapter Two)

No waking person is unaware for a moment of the uncertainty of outcomes in every enterprise. Although Heisenberg deduced his principle from the behavior of electrons in sub-atomic domains, and although he did not believe he had unsettled Newtonian rigidity from the big world of big things, his principle, and his actions "shake the foundations."

While our quotidian Newtonian world-view does not require modification due to the Heisenberg uncertainty principle, our ultimate knowledge about and ability to predict and understand the world does indeed need to take into account fundamental uncertainties.

Some of these uncertainties relate to the technical details of quantum mechanics, the rules that guide, or that permit the fuzzy existence of electrons. The domain of electrons seems to be intrinsically blurry, uncertain, incapable of being deterministically pinned down. But if our world is built up from atoms that contain electrons, and if electrons are unpredictable and uncertain, isn't our gross world built on an unreliable platform? Electrons are often described as "smeared out," or "like clouds," with areas of density but also with some internal space. After Heisenberg, we understand that "electron" is a *concept* that we use to express the probability of finding a tiny amount of matter, energy, and electronic charge, in an area. This idea, that the world is built on the probability of occurrence, rather than on the durability of "things," is the key to particle physics, quantum mechanics, and our understanding of the bedrock of matter and the world. After listening to the voice of Heisenberg, we know that some events are likely, some events are unlikely, and nothing is "certain."

Truth is a probability statement. When we define what we know we need to also clarify how we have come to know it. We always know what we know based on something, and that basis may contain hidden viewpoints and biases.

We also are reminded once again how what had once seemed to be apparently unimpeachable and highly reliable scientific laws, like Newtonian physics, may only *appear* to be accurate, and may even be simply untrue if we explore different compartments of reality. Newton's great bedrock principle, that the same laws affect Earth and heaven, that the universe is one domain, no longer seems true. The extension of earthly laws did make it possible for us to understand orbits of other planets, but Newtonian laws cease to rule the world when we observe the very small spaces where sub-atomic particles like electrons reside. But then again, Newtonian physics correctly predicts that basketballs never spontaneously leap up into the air to score two points on their own.

How many domains of reality are there?

If we exist, as Einstein says, in a context of an intrinsically uncontactable beyond, might not the pinnacle geniuses of Heisenberg and Gödel themselves with their apparent irrefutable "certainties of uncertainty" some day be scaled, and overturned? Isn't "uncertainty" itself uncertain?

Can we gaze upon the sunrise from any other position than tentativeness, and wonder? Can we ever feel we are firmly ensconced in the immutable fact? Aren't we always "knowing" with a mind the nature of which itself is in question? Do you want to base the meanings and purposes of your life upon discoveries, scientific laws, conclusions, which in the final analysis are the best we have but which are still subject to being over-ruled some day? Isn't all truth subjective? Doesn't the uncertainty principle and quantum mechanics imply that there is no such thing as "objectivity"?

These head-banging questions were tackled by a physicist and philosopher who wrote *On Physics and Philosophy*.

The French nuclear physicist and philosopher Bernard d'Espagnat, who is the preeminent philosopher of quantum uncertainty, and who, like Paul Davies, was awarded the Templeton Prize, writes that the subjective mind of the scientist, which some theorists claim is always distorting and subtly personalizing any experiment, is in itself an objective emergence of the world, which has created subjectivity. What appears "subjective" about our mind is after all a product of the laws of the universe. Can the universe be said to have created subjectivity? The subjective properties of mind may themselves be emergent properties of the objective world. The objective world creates the mind with its subjectivity.

When we try to understand, are we distorting due to our subjectivity, or are we seeing the world (as Morowitz might say) with the very mind that is embedded in the matrix of the world? Or are we distorting with a subjectivity which is itself an expression of an intrinsic uncertainty and unknowability that is fundamental to all things, not just our minds?

D'Espagnat has written that any reality we construct is an ongoing approximation, an asymptote approaching, but never reaching, a deeper, "veiled reality." He clearly states that he is not denying the valuable insights that are based on centuries of careful evidence and powerful logic that constitute science. He himself lived his life as a distinguished physicist. Not all statements have equal degrees of data, thought, and truth-value. (See also Chapter Sixteen). Good scientific research, he says, tends towards the real. But as close as we get, as near as we tend, a veil remains between any assertion we can make and an eternal truth. Aren't degrees of uncertainty and unknowability our fate? Aren't thoughtful people today left in wonder about the world beyond the *veil* of our own mind, historical era, sensations, perceptions, beliefs, thoughts, and sources of information?

Can we accept this, our plight, by building civilizations not upon beliefs, but upon ongoing wonder? Is religion by definition required to provide answers? Can't there be an organized religion of questions and wonder? Theologian A.J. Heschel wrote about his deepest request from God: "I didn't ask for success. I asked for wonder."

The added information, multiplicity of perspective, and uncertainty in the modern mind is a blessing and an invitation. We no longer live in fortresses. Is doubt our enemy, or our dearest friend?

Chapter Eight

Most Beautiful and Most Wonderful

Wonder is not always benighted. Great voices of wonder include cautious pessimists, as we have seen represented by two Nobel Prize winners, Stephen Weinberg and Jacques Monod, who were quoted in Chapter One. Maybe part of the problem is that we have been standing too far away from ourselves, poking at invisible electrons and irresolvable philosophies. We need to consider life. Life has added dimensionality and provokes our wonder in a more visceral way than does abstract science. What can life on Earth do for our sense of wonder? Great evolutionary biologists have added their voices to the question of the wonder of life.

The nature of our mind remains elusive because our biological essence itself remains "an irreducible mystery." Are we unique? When we look up into the sky at night, are we facing, across the cosmos, friends, enemies, inscrutable otherness, or nothingness? Our minds emerge from the matrix of our bodies, which scientists like Lewis Thomas and Carl Sagan viewed as lucky happenstance. If we are just a quirky happening in the roll of the cosmic dice, that might place our "mind" and our "truths" in a very different light than if life is an *inevitable* expression of the universe and its varying domains of laws. Are our wondering minds weirdo accidents or destined creations? Our own science, our own subjectivity, our own uncertainties may appear different to us depending upon whether we feel we are *intrinsic* to, or merely an *afterthought* of, the cosmos. But after reading Sagan's description in Chapter Three of one sperm cell out of millions winning the lottery to form every one of us, how can anyone refute that life is mere luck? Someone has.

Christian De Duve is one of those meticulous scientists, who avoided speculation and focused his life on experimental facts. He was awarded the Nobel Prize in medicine for his discoveries about the little lifelike forms, lysosomes, that exist even within, and smaller than, cells. Then, defying William Wordsworth's belief and Rachel Carson's worry that wonder decays with age, the Laureate let his mind soar.

He wrote in his beautiful book, *Vital Dust*, "Organic carbon compounds are everywhere. They make up twenty percent of interstellar dust, and interstellar dust makes up 0.1 percent of galactic matter." A tenth of one percent is a small fraction, but a gigantic total amount if you are talking about all of the matter in the universe. De Duve continues: " In this organic cloud, which pervades the universe, life is almost bound to arise... trillions of foci of past, present, or future life exists...biospheres coast through space on trillions of planets, channeling matter and energy into the creative fluxes of evolution. In whatever direction we turn our eyes... there is a life out there, somewhere...the Earth is part, together with trillions of other Earth-like bodies...of a cosmic cloud of vital dust...the universe *is* life." De Duve states as his conclusion to a lifetime of biomedical eminence: "Life is likely, reproducible, and commonplace."

Christian De Duve wants to overcome Jacques Monod's "alone in an unfeeling universe," and Steven Weinberg's "pointlessness." He does not exactly partake of Lewis Thomas" "contented dazzlement" or Thomas Wolfe's "new magic." He is not exactly saying that life is a *product* of the universe. Just as Einstein saw reason "incarnate in existence," De Duve says, "The universe *is* life." For him, life is in the universe as the fetus is in the mother. The universe is alive and biological life is *a destined and inevitable expression of it.*

Is it correct to ask the question along with most scientists, about how life arose on Earth? Or should we assume with Christian De Duve that the universe intrinsically contains life, as Morowitz assumes the universe to be intrinsically noetic? Is our inhabited planet a quirky cosmic creation, or an inevitable expression of a universe whose information arrows drive matter towards vitality and consciousness? According to De Duve, the laws of physics and chemistry focus the inorganic matter of the world into life in the same way that a hillside forces water to run downhill. True, there can be many different rivulets and streams running off the same mountain, and, true, chance may be involved in determining the specific channel of each stream, but all the water will be guided by the laws of the universe, in this case, gravity, to flow down. Similarly, the universe conducts its immeasurably vast cloud of vital dust into the chemical combinations which start life, and which then evolve into more complex life.

For De Duve, the equivalent to the downhill force of gravity on organic chemicals starts with electron exclusion behavior. Because, as Pauli pointed out to us, electrons avoid crashing and fusing, matter in the form of elements can get together in stable chemicals. Morowitz has written: "Due to exclusion, entities act like chemicals." And due to chemicals, the approximately one hundred elements can *recombine* into the millions of molecules that form our diverse world. There is a downhill flow: exclusion organizes electrons in atoms, which organize chemical interactions, which organize life. Because electron clouds contain awareness, chemicals relate and create in myriad forms that converge as life.

Can De Duve be right? Doesn't he seem to be in conflict with Lewis Thomas and Carl Sagan's world of luck? Why does no less a person than the legendary cosmologist, Stephen Hawking, say that life is a rare event? If De Duve sees life as an intrinsic property driven downhill by the universe, then why does that other great life scientist of the twentieth century, Steven Jay Gould, see life as improbable luck? Life, according to Gould, is enduringly and predominantly bacterial, while our upright posture and thought are transient and insignificant quirks. Life, according to Gould, is unlikely, and when it occurs, it is predominantly interesting but not terribly tall or smart. Gould wrote, "Life, is a glorious accident." The story of life on Earth does not aim at or converge on humanity. Anthropologist Loren Eiseley wrote in *The Firmament of Time* that the shaping force in evolution requires "…the irrationality and waste of extinction."

How could Harvard's reigning master of paleontology and the history of life, the late Steven Jay Gould, reach such emphatic and intransigent conclusions, about life on Earth being mere accident, given the fact that as a scientist, he saw, and he himself was a generator of, radical new ideas, revolutionary insights, reversals in one generation's world view? Could Gould actually believe that as life sciences evolve in the future, nothing will come along to shatter his own viewpoints? Did he believe his conclusions would last for 100 years, 1,000 years, 10,000 years? Who appointed him, what force elevated him so high, that he could look down upon the entire star-spangled cosmos with such conclusiveness? Yet if we ignore Steven Jay Gould's clear-headed science, rational processes, and evidence-based thinking, aren't we pandering to our own wishes, indulging in self-deception, living our lives under a cotton-blanket of fantasy beliefs? In clear, sophisticated logic, Steven Jay Gould has shown how probability theory alone accounts both for life on Earth, and its improbability elsewhere.

One of the most counterintuitive features of randomness is that it occasionally contains ordered sequences that appear by chance. For example, if you are flipping a coin, heads and tails will alternate randomly around the fifty-fifty probability, but occasionally a random "streak" of heads or tails may occur. An observer who only looked at the streak could easily misconstrue design, or directed sequence, within a statistically probable quirk within a longer, truly random series.

Given the billions of years of Earth evolution, and the zillions of bacterial and other primitive lives, isn't it possible, or probable, that complex life might briefly appear in a "streak"? Complex life is less than a billion years old on a much older planet in a much older universe; therefore there has been plenty of opportunity for random events to accidentally coalesce into a streak.

Steven Jay Gould's logic has been clarified by mathematical physicist Leonard Mlodinow, author of *The Drunkard's Walk*, who has said in a *Wall Street Journal* interview, "Random processes display periods of order…we find false meaning in them…looking for order in patterns has allowed us to understand the patterns of the universe, but we reject randomness due to our need for control…it is human nature to misperceive chance…extraordinary events can happen without extraordinary causes." Mlodinow concludes that there can be patterns inside randomness and randomness inside patterns. Doesn't it seem credible within the huge time periods and numbers involved that life is indeed accidental?

De Duve's ideas immediately resonate with us. They feel good. How wonderful to imagine that we have been destined to be born from a cosmos that has a direction, while cranky old Steven Jay Gould says we are mere accidental freaks. Are our emotions valid intuitions, or mere wish fulfillments?

Many other life scientists have argued against De Duve's "vital dust" worldview by pointing out the unbreachable distance between simple organic compounds that De Duve talks about, and life. All of life is based on cascades of complexities: enclosed cells, which store information in DNA, transcribe DNA information into proteins and metabolism, and replicate with complete copies of information. (Sections III, IV and V will elaborate this point in detail). Life often seems unbridgeably more complicated than mere organic compounds. The *interrelationships* within the living cell are what seem so inexplicable. "Vital dust" does not store information in sequences, translate information from one type of chemical to another, nor replicate itself. Can such abilities evolve from carbon compounds floating in clouds in space? Cosmic carbon dust and cellular life are not necessarily related. This is the argument that separates Christian De Duve and Steven Jay Gould. Vital dust and life may be like parent to child, or may be separate and only vaguely related phenomena.

Then again, within the context of a vast number of repetitions, improbable events become likely. If an unlikely event, which might occur only once in a million, is given a trillion opportunities, it becomes constrained towards the probable. So if trillions of bacteria lived on the early Earth for billions of years, unlikely chance events, like the development of multicellularity, become likely. That is how Steven Jay Gould accounts for what he considers to be the improbable and accidental emergence of human life, and how Christian De Duve accounts for common, simple, carbon compounds organizing into cell walls and informational macromolecules like DNA, upon which evolution can then work subsequent transformations. Probabilities lead Gould to view life as luck and De Duve to view life as inevitable. Both views derive from probability theory, which can be understood to mean that we are unique, or that we are common and probable. You can say that the huge numbers and long eons of trial and error made a rare lucky event happen one unique time, or you can say those numbers and eons made the lucky event constrained toward the probable, even destined.

There are few places on Earth where you can see a one billion year old rock. Such locations do exist in the Adirondack Mountains of the Northeastern U.S.A., or in the Canadian Shield, or on the triangle that forms central India, but they are unusual. Most of the time any stone, or bedrock exposed along a highway, or mountaintop that you see, is much less than a billion years old. Yet our own DNA is a copy of a copy of a copy, a soft thread in a cell that has been duplicated repeatedly and continuously for three to four billion years! *We are older than the rocks.* Some of the information that codes the molecules of which we consist is older than almost anything else we generally encounter on Earth. Our spool of DNA contains sequences that are older than the Atlantic and Pacific oceans, older than the individual continents. In my cells are informatic chemical sequences that chide the mighty Himalayas as children.

Life is not brief, has not been defeated by entropy, (which we will discuss in Section III, Chapter Thirty-Five) and is not easily accounted for as an accident. It has resolutely barreled on, charging over the chasms of eons. We say colloquially, "life force." Our invisibly thin DNA has won the marathon against gneiss, granite, and basalt. Doesn't this fullback sound more like a downhill cosmic force than like a chance tourist?

Then again, the Earth, the moon, and Jupiter, "the morning star," are older than we are. Are we visitors here for the holidays, or scions of an ancient house?

Would Charles Darwin have agreed with Christian De Duve or Steven Jay Gould?

Darwin, who is often associated in the popular mind with an impersonal and mechanical universe, nevertheless was a high druid of wonder. He spent nine years reveling in the intricacies of fossil barnacles! When he was severely ill and bedridden, he fixated on how the leaves and tendrils on the plants hanging in his bedroom responded to light, and there, as in so many other places, he discerned new operating principles of living beings. Did Darwin think of life as a cosmic force or as a one-shot deal? The always polite, always wonder-struck Englishman ended his opus *On the Origin of Species* with these lines: "...from so simple a beginning endless forms most beautiful and most wonderful have been, and are being evolved."

Chapter Nine

One Night in a Thousand Years

What exactly was that simple beginning? Darwin was referring to the origin of life on Earth. But there are other ways to think about origins. Like the Chinese puzzle, there are origins inside the origins.

The University of Chicago has produced some remarkable astronomers. One was Carl Sagan, whose cinematic and literary *Cosmos* could be called the National Anthem of wonder.

Human beings of the late 20th century were the first to have a sweeping, fact-based, world view, that incorporated atoms, energy, space, time and the heavens, in one embracing, interconnecting theory. Although the basic science of our cosmos is too extensive for any individual to master in detail, Carl Sagan managed to capture it in one image: "If you want to make an apple pie from scratch, you must first invent the universe." If you are planning to use atoms, if you are planning to use heat and oxygen and chemistry, you have to borrow your apples and crust from the universe itself, which made your basic materials and your cooking chemistry. As for our minds, Sagan quotes another poetic metaphor from the writing of neuroscientist Charles Sherrington: "the one hundred trillion connections in the human cortex form an enchanted loom of shifting harmonies." Sagan quoted as fact the words of science fiction writer, H.G. Wells: "Beings now latent in our thoughts and loins will reach out their hands among the stars."

Sagan's sense of "the beginning" is very distant from Darwin's. Darwin is talking about the Earth. Sagan's apple pie is as old as Werner Loewenstein's information arrows, like gravity and electromagnetism. Sagan's pie has ingredients which have stayed fresh for about fourteen billion years. (The Big Bang is variously dated to fifteen, fourteen, or thirteen and seven tenths billion years old by the various authors I quote in this Section. The current dating is 13.7 billion years.) Maybe the beginning of life on

Earth was not a beginning at all, but a continuation of more far- flung life. Are any other stars, like ours, life-giving suns? When we think about our origins, should we be thinking about Darwin's forms most beautiful, or Sagan's cosmic pie? That question brings us back to the creativity of stars.

Stars are associated with wonder not only because of their beauty, but because of their "otherness." Their very presence calls into question who, aside from ourselves, is present. Whether consciousness is a fundamental property of the universe, as Morowitz implies, or a unique, singular property of earthlings, as Gould believed, or an irresistible destiny of organic matter, as de Duve argued, we look up at the stars with mixed emotions, seeing them simultaneously as unknowably distant and as exquisitely familiar. They are among our oldest childhood friends, but we don't know them. They stir wonder that is aesthetic and existential.

Now our wonder is provoked by both the uncertain nature of scientific law, causality and chance, and also by the ambiguous meaning of our own life. Even through powerful telescopes, we do not yet have the ability to tell whether many stars are accompanied by potentially life-bearing planets. The several thousand exo-planets that we do in fact know about are based on surveys of only the closest region of our own galaxy, and have little bearing on the hundred billion other galaxies. The existence of life in other galaxies is not a problem amenable to current science.

Stars were gems of wonder for Rachel Carson, as they were for Carl Sagan, Walt Whitman, Rabindranath Tagore and almost everyone else. Stars are part of our shared, cross-cultural, universal human heritage. Rachel Carson describes going down to a headland on the bay of Maine, where "...the horizons are remote and distant rims on the edge of space... millions of stars blazed in darkness...the misty river of the Milky Way flowed across the sky...once or twice a meteor burned its way...It occurred to me that if this were a sight that could be seen only once in a century... this little headland would be thronged with spectators."

Carson must have absorbed Ralph Waldo Emerson's writing so deeply that she forgot she was paraphrasing his famous lines from the essay, "Nature." "If the stars should appear one night in a thousand years, how would men believe and adore..." Emerson's lines about the wonder of stars inspired the young Isaac Asimov to write what has been called "the best science fiction story of all time."

In "Nightfall" by Isaac Asimov, people who live on a planet that rotates around five suns only see the stars once in a thousand years, because almost all of the time they are rotating around one or another sun in day-light. In this fantasy, written in 1941, as Hitler was conquering Europe, the stars finally appear after a long prediction and a feared nightfall "...and in the blankness they appeared in countless numbers...of such beauty that the very leaves of the trees cried out in wonder." Although the leaves cried out in wonder, the humans in Asimov's story, like the humans in Europe in 1941, did not. Frightened by darkness and their first sight of stars, Asimov's fictional civilization progresses through panic, terror, prophecy, and superstition to frenzy and self-destruction.

Do the stars in fact really evoke wonder? Does everyone feel the way Rachel Carson and Ralph Waldo Emerson did? Is the cosmos soothing or horrifying to us? Is it a matter of what we actually see, or what we have been *taught* to believe? In the night sky, do we see "constellations," "Leo," "Greek Gods"; or do we see space, eternity, darkness; heaven, or the void? Would nightfall once in a thousand years evoke belief and adoration as Emerson said, or immolation by flames, as Asimov implies? In Asimov's view "there remain the children below six, to whom the world as a whole is too new and strange to be frightened of the Stars and Darkness...the stars would just be another item in an already surprising world."

If Rachel Carson takes her nephew, Roger, down the balsam-scented slopes of the Maine coast at night to see the stars, will he be imprinted naturally to wonder, or will he need to be taught to wonder? Will he need to be didactically de-conditioned from awe and dread? Are stars friendly lights blinking at us, or are they horrid furnaces of a consuming nothing-ness? Is this what the authors of *The Bhagavad Gita* had in mind? The nature of scientific law and chance leaves us wondering, and so does the nature of life. Now the cosmos and our reaction to it also provide wonder.

It is easy enough to claim that stars evoke wonder when you inter-pret their lights in the context of your comforting beliefs. But what if the stars are only fragments of an impersonal, all-consuming, all-destroying explosion? Don't stars, when viewed without preconception, also provoke a feeling of "pointlessness" and "alone in an unfeeling universe?"

Before Carl Sagan, another University of Chicago PhD in astronomy had already been awarded to Edwin Hubble, the scientist who may be said to have created the modern cosmos that Sagan popularized. Hubble's work at Mt. Wilson in the 1920's and 30's, with what was then the best tele-scope in the world, established conclusively that the hazy "nebulae" were other galaxies. Hubble, modestly using the term, "we," (probably to refer to the telescope and himself) said at Yale, (as quoted in *The Realm of the Nebulae*) that his work "...enlarged the domain of positive knowledge a thousand-million fold." After Hubble, humans went from imagining ourselves to be in a star-studded world, to awareness of a vast, multi-galactic space. With the discovery of Quasars, the Hubble cosmos took another incomprehensibly gigantic leap in size. (This will be discussed in Chapter Fifteen). Hubble's revolution in our sense of the world and our place in it, can only be compared to that of Copernicus, who ejected us from center field. Hubble flung us out of our visible galaxy into the cosmos of billions of galaxies. But Hubble then went on to something grander.

By using a situational property of light rays, called a "red shift," that enables astronomers to measure the distance that light has traveled (see also Chapter Fifteen), Edwin Hubble proved that the universe was expanding. His discovery was the catalyst for the Big Bang (a term he himself never used) since, if it is getting bigger all the time, the universe must once have been a lot smaller. At the height of his remarkable fame and glory, the much-adored Albert Einstein made a pilgrimage to Mt. Wilson, and, ever honest and modest, announced to the press, and to the world, that his own previous belief in a static universe was wrong, as proven by the observations of Edwin Hubble (cosmic expansion will be discussed in more detail in Chapter Seventy-Seven).

As radically as Einstein and Hubble each transformed our psychological world, both remained within the framework created by Sir Isaac Newton. Part of Newton's eminence rests on his articulation of the principle of one world, where all laws operate equally everywhere, as we have seen. The heavens are not a separable creation with unique laws. The visible world is assumed to be uniform throughout time and across space. By applying the rule of unity and uniformity, Newton could explain the rotations of heavenly planets by earthly laws of gravity, and his laws still operate in our understanding of the sky. (The quantum revolution of Planck, Einstein, Pauli, Heisenberg, et. al. applied only to the domain of the very small and invisible). Einstein never slackened in his reverence for Newton and used the same principles in his relativity theories that deal with the large-scale cosmos as a whole. Some laws, like the constancy of the speed of light, seem universal and cosmic. But Philip W. Anderson unintentionally spearheaded a new revolution, as radical as that of Copernicus, Einstein, Hubble...or even more so.

Chapter Ten

"More is Different"

When we listen to voices of wonder rising up from the second half of the twentieth century, we hear about new information, such as the existence of galaxies or quasars, and we hear about new domains of natural law, such as the small quantum world, or the early "Big Bang" universe in which the weak and the electromagnetic forces were still unified. We also hear new ways of thinking. We not only have new information, we have new pathways to insight. Our concept of "cause," whether implicit or explicit, determines how we see the world, and how often and how deeply we experience a sense of wonder.

The idea of "cause," how things influence each other, how forces operate, has been reborn. Simple cause-and-effect thinking has been expanded. We now are capable of relating events in more diverse and more realistic ways. Causality is less like a line of force, and more like a web strung throughout events. The most piquing questions are not only, "What does the world consist of?" but also, "How does the world hang together, connect, communicate, and relate?" How do things and events influence each other?

There are many types of stitches in the fabric of the world. There are many types and degrees of cause. The world develops and transforms into many things in many ways, and recognition of the variability within formation erases old assumptions and conclusions. Although this recognition dates back at least to ancient India, it took center stage in late twentieth century science due to Philip Anderson, his aphorism, "more is different," and the inspired scientists who followed him. The revolution within the revolution was not increased information from telescopes and electron microscopes, but new understanding of relational influences that create form in the fleeting world.

Anderson was a physicist at Bell Laboratories in New Jersey in the 1960's. Disturbed that research money was being overly directed towards conservative, traditional paradigms, instead of innovation (such a pragmatic, banal, fiscal concern!) and based on his own original probes into uncharted physics, he wrote an article in *Science* in 1972, titled, "More is Different." Anderson argued in a pithy, metaphorical essay that not only do the mosaics of the world's particles change, but that new laws also emerge within the universe. "At each level of complexity entirely new properties appear...a shift from quantitative to qualitative differentiation takes place...At each stage entirely new laws, concepts, and generalizations are necessary."

This position represents a radical break from that of Newton, Einstein or even Genesis. The universe has always been understood to be creating new things—mammals where there was once only bacteria, planets where there were once only stars. Anderson was arguing that not just the stuff of the world can be reorganized, but that the very laws by which this can be accomplished may also be reborn. Today's laws are not the laws of yesterday. Just as new things can be created, there can be new processes of creation. Conversely, new things themselves may catalyze new fundamental governing properties within and around them. Stuff can be obedient to laws, but new arrangements of stuff can bring new laws out of the backstage, behind or deeper within the universe. The new laws are consistent with underlying laws. They are not *de novo* disconnections. They are not arbitrary products of capricious will. Yet new laws emerge to elaborate an increasingly complex universe. Not just the numbers and sizes of things become more complex, but new relationships emerge within the cosmos.

Anderson received the Nobel Prize in Physics in 1977, primarily for studying disorder in a manner that foreshadowed and facilitated electronic switches in computers, and today the idea of "emergence" is widely pursued in physics and biology. For example, under the conditions of extreme compression and heat believed to have existed during the early stages of the Big Bang, different laws may have guided the embryonic forms of energy and matter than the laws of physics as we know them today in the context of our huge, cold, expanding universe. The laws of physics, which appear to surround us, and to guide our universe, may in fact be local, or temporary, or "phase specific." In Anderson's words, "matter will undergo mathematically sharp, singular, 'phase transitions.'" Our laws may exist only within the context of certain zones of temperature and pressure. The world can tip over into a radically new and different state. It is as if new fundamental relationships condense out from the fluid universe.

For example, the "weak force," which appears to be a fundamental force of the universe that acts inside of atoms, and that behaves apparently very differently from electromagnetism, and that therefore has its own name, and its own mathematical formulas, may actually be a form of electromagnetism. Today scientists believe that the "weak force" was unified with, or was the same thing as, electromagnetism in the early seconds of the universe. As the temperature and pressure of the early, hot, compressed universe expanded and cooled, the "weak force" separated from electromagnetism and now appears different in our older universe. This was the conclusion that led to the 1979 Nobel Prize in physics being given to two graduates of the Bronx High School of Science, Sheldon Glashow and Steven Weinberg ("pointless," Chapter One), and to the British scientist of Pakistani origin, Abdus Salam.

Similarly, biological laws cannot contradict laws in physics and chemistry, but may also emerge uniquely to govern the biotic domain. Natural selection, the guiding hand in evolution as described by Darwin and Wallace, may be an example. Natural selection operates on and guides life in evolution, but has no role in physics. It is a law that has emerged with life. The laws that govern life may be unique to life itself and not reducible to their underlying physics and chemistry. On the one hand biological laws are not entirely independent from and can never be opposed to physics and chemistry. On the other hand, biological laws operate in spheres that physics and chemistry *do not constrain* and where biology can manifest its own new phenomena guided by its own emergent laws.

The laws of life are surrounded by and contained within physics and chemistry the way horses in a corral are contained by the fence. The horses cannot move outside of the fence, but within the corral their motions are not predictable by knowledge of the boundaries that the fence imposes. The horses, like most of life phenomena, reveal zones of deterministic limitation, (the fence) and zones of living freedom. Inside the fence, the horses follow horse-specific laws, such as deference to the dominant stallion. Horse behavior is partly a product of molecular, cellular physics and chemistry (that make the cells in their bodies), partly a product of mammalian evolution, (that make their gross, horse shape), partly a product of their human-dominated environment, (the fence) and partly horsing around.

Even within the so-called "hard sciences" of physics and chemistry there may also be firm limits surrounding fluctuant zones of emergent, creative, newness. Anderson's original paper showed how even temperature emerges only when particles exist at least as two. A single gas atom has no temperature. Temperature emerges when atoms *interact*. The heat of atoms in collision is a phenomenon that requires more than one atom. One atom has no heat. Many atoms together in an enclosed space generate a new phenomenon called heat. Even with "inert" matter, entirely new phenomena appear in the world with the addition of new particles. Heat emerges into the universe as a form of *relationship between two things*.

The laws of the universe may emerge from relationships. Before the relationships exist, the laws which will govern those emergent relationships also do not exist, or are latent, or are as yet unknowable. As we saw earlier (Chapter Five), Morowitz describes the Pauli exclusion principle as being an emergent property of electrons in *relationship* to each other. A single electron alone in the universe, not incorporated in an atom, but independently riding in pre-atomic plasma, cannot manifest "exclusion."

Even such impersonal, mechanical physical phenomena as heat or electron exclusion come into being through interaction, contact, "togetherness."

Another example of emergence and creativity that speaks to the uniqueness of life has to do with what scientists, like Anderson, call "symmetry breaking." Complex organic molecules of life often can exist in three-dimensional shapes that differ from each other in their spatial relationships, but which act identically in chemical processes. These different looking but similarly acting forms of the same chemical differ like my right and left hands. Because these different handedness shapes, called "enantiomers," behave identically, there is no reason for nature to favor one over the other. They look different but they behave the same. When chemists synthesize these types of molecules in a laboratory, two shapes spontaneously appear in approximately equal numbers. The laws in the zones of physics and chemistry spontaneously lead to a fifty percent to fifty percent mixture that has "symmetry," equality between the two forms of the chemical.

But in the transition from molecules in non-living systems into biological molecules in cells, the symmetry is broken. Life chooses one spatial shape of the molecule, and the other shape of the molecule is eliminated even though there is no physical or chemical basis for this selection. The selection seems to be based on new laws of biology that have no meaning within the realm of physics and chemistry.

Yet these new biological, symmetry-breaking selection laws are the same in all of life from bacterial cells to human cells. These new laws span the entire tree of life. The symmetry-breaking selection laws are a property of life that emerges with life, and that is consistent with all of life, but is not guided by the laws of atoms and rocks. So, for example, glucose sugar molecules, which have two equal and opposite forms that can work perfectly well together or apart in chemical systems, nevertheless are found in *life* systems only as *one* hand, one right-handed form, while the other handedness is *systematically deleted from every living cell on planet Earth*! Life has risen up out of the constitution of physics and chemistry and has legislated something new for no immediately clear reason. Life has its own laws, it own domain.

The apparently old laws that guide life may be second-generation immigrants, (like the "weak force" form of electromagnetism which also emerged through symmetry breaking) and we ourselves may be products of *layers* of laws that have only recently (say three or four billion years ago) arrived, as part of the selections made by biological systems from within the large domains of physics and chemistry.

Is there anything constant, immutable, eternal? Are scientific laws fixed into the order of everything like "the mind of God" or are they too changelings, and if so, of what? Before emergent laws emerge do they exist in waiting, like relief pitchers in the dugout waiting to be called into the closing innings, or are they entirely non-existent until they emerge?

Is there a "mind of God," or an "emerging mind"? Is the universe coded, or emerging anew out of itself? Is there "a mind of God," or "minds of God," or "minds of Gods"? Can the universe give birth to an ever-new universe?

Even when we consider the laws, which science extracts from the apparent regularities of the world, to be fixed, eternal, and universal, how were they chosen? Who/what arranged them in this way? (This is also discussed in Section III, Chapter Forty-One). Why not have gravity be a little stronger? Is there a law behind or within them, a single basic law, as Einstein believed and struggled unsuccessfully for decades to find? And if so, where did that core, singular, mother of all laws come from? And working our minds downward in the other direction, once the world and its regularized processes exist, do they self–interact to engender entirely new laws never before existent?

In Anderson's words: "The whole becomes not only more than but very different from the sum of its parts." In the history of the universe, alongside new combinations of old things, entirely new systems of components and governing laws emerge which are unique and inerasable. The world has been in the process of irrevocable self-transformation. The universe emerges out of the universe: this is the realization called "emergence."

Stuart Kauffman, a leader in emergence and complexity studies, has written in *Reinventing the Sacred*, "The whole acts on the parts as much as the parts act on the whole." The totality impacts downward as complex worlds, like planet Earth, grow and change. Plants create an ecosystem, and the ecosystem then regulates relationships between its participating plants. First the parts become the whole, and once the whole exists, it acts backward and downward to reshape the parts that initially shaped it. This is the meaning of "more is different." The original parts become transformed and cannot resume their earlier configurations, so that *history*, the history of life, the history of our planet, is *symmetry-breaking and irreversible*. You can't get back to the origins, which now, in the context of a new whole, have been smudged and transformed. You can't go home again. The eggshell is broken and the bird has been born.

We always live in a new world.

For example, long, slow evolution on Earth gradually produced photosynthetic cells which created the molecular oxygen that now seems so intrinsic to our planet. (Oxygen, like all large elements, is created by nuclear fusion in stars, but breathable oxygen, O_2, is created by photosynthesis.) The early cells employed chemical elements drawn from Earth rocks, like iron and phosphorus. Once the biological molecular oxygen came to exist in significant amounts, the elements within the rocks often changed upon contact. Some of the elemental rock became "oxidized." The oxidized elements in the rocks behaved differently than the non-oxidized elements that used to be predominant. New things emerged from a basement layer and the new things changed the substrate from which they themselves had arisen. Furthermore, atmospheric oxygen in the form of ozone reduces ultraviolet radiation and thereby transforms the habitability of our planet. Our planet has reorganized itself "top down." Created things created new things, and the planet has been fundamentally transformed, not only within its living layers, but even in its mineral nature. Planet Earth today is very different than the planet that gave rise to atmospheric oxygen billions of years ago. Not only have new things been added, but old things, like freestanding unoxidized iron, have been erased. (See also Chapters Fifty-One, Fifty-Two, and Sixty-Nine).

Similarly, Freud emphasized how parents mold, help or harm their children. Today we also recognize how the offspring influence their parents. Causality is not entirely linear. The parents create the family, and the family system partly recreates the parents. The whole family reshapes the parts of the family that gave birth to it.

In the emerging universe, as parts evolve, they help to shape the whole, and as the whole evolves, it helps to shape the parts, so that creation is self-regenerating. Kaufmann's phrasing echoes what we heard in Indian art and philosophy (in Chapter Two), how everything must be understood in the context of its historical and contemporary relationships. Laws, like animals, may propagate into previously undefined terrain. As Kauffman phrases it, "the creativity in the universe is tied to the explosion into the adjacent possible." Wherever there is opportunity, there can be emergence.

The concept of "complexification" applies to laws interacting with laws (not just with things) to create new laws. "Emergence" applies not only to new appearances, but *new regimes of lawful complexity*.

According to the type of thinking of an Anderson or a Kauffman, the universe may be lawfully creative, historically singular, and *unpredictable*, derived from, but not accounted for, by its initial information state, because information and order themselves may be developmental and self amplifying. *The products of the initial information state may circle back and amend its original state.* Then the initial information state is gone, changed, forever transformed. *The first cause is not the source of all cause.* Causality has children who grow up to run the household. The initial information state has undergone regime change.

The original information state that was present in the early moments of this current expansion of the universe may not account for, nor predict, the totality of the universe today, because the current information state arises from *interaction, compounding, complexification*, and the *emergence* of new laws.

The world has not been printed out from an initial code.

It is the world that has created a new world.

Chapter Eleven

Knowing How Way Leads on to Way

To some degree, we have left Newton and Einstein behind. Of course they discovered enduring insights in certain domains. But when we look around the world, and wonder how it has come to be, when we wonder how we ourselves have taken shape, we now find ourselves in an inventive self-renewing world. Things keep changing, and the laws by which change occurs keep changing.

Stimulated by the questions implicit in science and causal thinking, we find ourselves wondering how the very laws of science themselves emerged. We can envision the universe itself evolving, developing, emerging out of and transforming itself. We should not be surprised to find that history and time are creative and constructive lawgivers, who have tinkered with the governing laws.

Not only can new laws emerge from complex interactions of phenomena following the original laws, (think of "heat" and "exclusion") but particular *paths*, once taken, may attain the status of a law. We are now adding to "cause," and to "top-down cause," another way to shape the world, "path-dependent cause." This happens when two paths may both be consistent with the laws, but once the left or right-hand turn is made, there may be no going back, and the subsequent unraveling of the world is contingent upon the choice that becomes "frozen" into history, a "frozen accident," or "broken symmetry." Once one path has been taken, and living forms come to embody it, then subsequent life becomes "path-dependent." Life and the universe have radical transformers, but they also have a fixed grid of conservative, old-boys networks. Once something is in place, it may well hold sway. The history of our planet and its life forms is unreturning or, as Alfred Russel Wallace, Darwin's cofounder of the understanding of natural selection, called it, "indefinite departure."

The world seems full of examples of "frozen accidents" that determined subsequent paths. A cell in your liver cannot suddenly decide to become a brain cell. Though it has enough DNA for that, it is no longer a stem cell, and has gone down the path of liver cell DNA expression. Similarly, the biosphere of planet Earth, which once creatively amplified dinosaurs, can no longer induce a tyrannosaurus to stalk among us. Earth has gone down a path that has left the dinosaurs behind. Although every part of their living bodies was consistent with the laws of life, a historical turn was taken, and the path of life now excludes them. What is once done, may well be done forever. Time itself may be a legislator, a Caesar who fixes laws.

The moon and its influence upon us may be another example of path-determined lawfulness. The moon consists of elements that occur in percentages similar to those within our own Earth mantle. This chemical similarity cannot be taken for granted. It is very unlikely that such a commonality evolved separately in two different heavenly bodies. Other planets and moons in our solar system, for example, Jupiter, Saturn, or their moons, have a chemical composition entirely unlike planet Earth. Therefore, scientists hypothesize that the moon is actually a piece of our Earth which was split off in an ancient collision four to six billion years ago. (The moon may also contain material from the impact object that split it off.) The moon's rotational and gravitational effects now help to stabilize the Earth in space, and this stability in turn has been helpful for the slow development of life over billions of years, which otherwise could well have been disrupted from time to time by planetary instability. Is it possible that a chance collision with an asteroid or a planet, some four billion years ago, which first broke off the moon chunk which then in turn helped to stabilize our planetary mechanics, was a happenstance without which we would not have evolved into existence? We cannot say that such a collision was caused. It must have been a very rare accident. (Lesser collisions with asteroids have been somewhat more frequent as the various known impact craters reveal). Such a massive collision would seem to have been a random, one-shot, uncaused event. Once it had happened, however, scientific laws moved down a path. The path was determined partly by natural laws and partly by a path-determining accident that got "frozen" into the subsequent history of Earth.

The concept of path-determinism is considered to be the probable explanation for something we puzzled over in the last chapter, why two equally distributed handedness of glucose exists in chemistry, but why only the right handed form can be found in living cells. Once one of those sugars had been selected, quite probably by accident, the living systems of cells could no longer function if the other form were inserted. If you are using your right hand to fix something, you cannot decide to protect yourself by using a left-handed glove. Once everything in your cells and in your whole body has been tooled to fit one design of sugar, you cannot arbitrarily flip over to the other design. If you drive a Toyota, you cannot use Chevrolet parts. So the phase specific laws of biology, which emerge from but become independent from physics and chemistry, may partly rest on path dependent causality, frozen accidents, a mix of happenstance and subsequent fixation.

Living beings, and the biological laws within them, throw us back into haunting questions about the fundamental operating principles within our own bodies and minds. Causality and chance are certainly not the same, but they aren't opposites either. Our existence as living things, as mammals on Earth, derives from fixed laws like gravity, happenstance like orbital smash-ups, and resilience, opportunism, and adaptation in the wake of events.

An adult does not exist without a pre-existing fetus. But an adult is not just a bigger or more complicated fetus. An adult starts with the conception of a fetus, but as we develop both in the uterus and after birth, we gradually become something that is radically new and different. Although our DNA is an important determining factor, so also are other factors such as intrauterine environment (did the mother smoke? drink alcohol?), our parents' personalities, our schools, culture, etc. History irrevocably enters in. Original causes like DNA set a direction for the path. This fetus will be a person, not a turtle. But historical situational factors also set in. This person will be brain damaged due to lead paint in his mother's apartment, and this other person will be funneled by opportunity and education towards an Ivy League degree. An adult is exactly derived from, and completely different than the fetus. Historical events get frozen into the trajectory of time, like Carl Sagan's victorious sperm cell, and Thomas Wolfe's Cretan love affair (Chapter Three). If by chance you are exposed to high levels of lead paint as a child, your intellectual development will be constricted, regardless of the original information state in your DNA.

Human beings seem to exemplify the concept of path-determined emergence. How can all the decisions that will be made today by all seven billion of us be determined exactly by physics and chemistry? Clearly, Einstein's determinism, which sought to derive everything from one mother law, or from a small nest of powerful original laws, does not account for the free movement of horses even within their corral, and does not account for the complexity of human culture in its seven billion-fold interactions. Instead, as life forms emerge and evolve, new things, like bacteria and people, and new governing properties among those things also have appeared in the world. I cannot blame bad decisions, like forgetting to take my raincoat, on the initial information arrows issuing from the "Big Bang" fourteen billion years ago. But as we will see in Sections III, IV and V, the essence of who I am does indeed pay obeisance to ancient and original laws.

Liberals and conservatives share governance in the parliament of my being. Old laws and new amendments are both on the books. Nonreversible nodes in time cue up all the subsequent events.

The concept of path dependence helps to melt away the false dichotomy between law and chance. Chance events, like asteroid impacts on Earth creating a moon, like sugar selection by cells, can morph into law. That means that over billions of years, a lot of usurpers have now become rulers. (See Chapter Twenty-Seven). We follow cosmic law, historical events, and fixed traditions.

The world we are born from is causal but not all caused; it is lawful but not deterministic. It has revolutions and stable republics. The world is neither formatted from a pre-existing design, nor is it entirely unpredictable. "Cause" is an overarching, general term, that refers to the myriad ways in which matter, energy, and information interact, relate, touch, constrain, influence, enhance, excite, or limit each other. There are strong grips, deep channels, glancing blows, and whispers.

Mind, with its ability to represent situations that are not immediately present, enabled pre-planning and anticipation (Chapter Six). Of all phenomena, mind seems least reducible to fixed, original causality, and seems most consistent with gradual emergence through biological complexification. Mind anticipates, that is, imagines de novo within brain-space, and then seems to select. Something is imagined, then chosen, and then emerges in action. Mind seems both path-determined, emergent, and an example of top-down processes. We human beings, who are highly organized bags of earth-derived chemicals, have massively rearranged and changed the

Earth itself. Our minds and cultures emerged from chemistry and biology and then reorganized Earth, top down. Once we emerged from within the lawful zones of physics, chemistry and biology, we humans then influenced from the top down the systems that had given us birth in the first place. We rearranged the platform from which we arose. We are a product of very long evolutionary paths, that survived many quirks and extinctions, and once we got on stage, we re-arranged the whole theatre. Einstein might wish for an old Monarchial god who does not play dice, but we are best accounted for by concepts like path-dependence and top-down causality.

The plow, the chain saw, the bulldozer, and the factory fishing trawler have scraped and reshaped the mother-load of our biological cradle, Earth. We evoked wheat and corn out of the genes of primitive grains, and now cannot survive without them. Concepts like path-dependence, and top-down causality, expand our thinking beyond the more limited physical determinism that guided Einstein's thoughts. Einstein, like all of us, was a product of his historical era. The most modestly endowed among us are now the inheritors of late twentieth century scientific explanations that carry us into more satisfying and more intricate answers than Einstein was able to formulate. We are lucky to live in this era of time and we are reminded of the caution that is necessary when stating any so-called conclusion.

Today, when we think about ourselves, we can see that we are consistent with, but not a mere printout of, scientific laws. Amplifications, emergences, choices, accidents are woven into the crossing threads of deterministic forces. The world is lawful but not predictable. There is order, opportunity, flexibility, chance, growth, development, creation. Newtonian physics and Lewis Thomas's dazzlement of surprise co-exist. In the past, determinism got confused with causality. The universe sees no contradiction in containing domains of law and domains of luck. There are Newton's laws for the rotation of planets, and there are freaky asteroid collisions with Earth. (Chapter Twenty-Seven will expand the importance of path-dependent cause).

Robert Frost's poetry, like that of Walt Whitman, was attuned to the science of its era. His poem, "The Road Not Taken" says:

"…I kept the first [road] for another day!
Yet knowing how way leads on to way,
I doubted if I should ever come back."

Robert Frost had just one of the two possible rotations of sugar molecules in his cells, and enjoyed life on a planet where a stabilizing moon, cracked off in an accident, co-nurtured long development of life. It is one thing for us to enjoy Frost's piquant phrases. It is a source of wonder when we consider how a long slow brew of rocks, elements, chemicals, cells, reptiles, lemurs and such, compiled the atoms of the universe into Robert Frost. He chose one path in the woods, and doubted if he could ever get back to the road not taken. How many similar alternatives had been taken before suns, planets, rocks, algae, bacteria, dinosaurs, and apes shaped atoms into the path of becoming Robert Frost?

Chapter Twelve

The Sound of Twenty-Six Letters

In the late twentieth century, science not only expanded our information, but expanded our concepts for understanding the information we had gathered. The previous, apparent paradox that seemed to exist between determinism and randomness, between causality and freedom, is now best understood by using the concept of "deterministic chaos." Along with emergence, top-down causality, and path dependence, deterministic chaos provides us with yet another tool to open the door wider than Einstein was able to do. We can think about ourselves and our world with greater flexibility, appreciation, and curiosity than when we were stuck on simple ideas like "luck" or "cause"

Deterministic chaos, as defined by Werner Loewenstein, for example, in his book, *The Touchstone of Life,* (see also Chapters Three and Sixty) tells us that lawfulness can predict unpredictability, that order and cause can produce disorder! Here is how it works: If initial conditions include enough combinations and possibilities, then the combinations and permutations of the origin may expand beyond prediction and calculation. In this case, there is law behind and within the situation, but the possible number of rearrangements of the original elements becomes too large for any single outcome to be probable. The situation follows law, yet is unpredictable. Information therefore can produce chaos. This is different than randomness which is merely the absence of information. Instead, we are dealing with information in the universe which codes, predicts, and determines states of unpredictability, variability, and creativity. Indeterminacy is not always the opposite of order and information. Unpredictable events may emerge from lawfully structured precursors.

Harold Morowitz, in his book *The Emergence of Everything* (see also Chapter Five), has described deterministic chaos like this: Even within the confinement of rules, unexpected results emerge, due to variable starting conditions, variable boundaries, and the explosion of interacting combinations, as well as the pruning back of variability once a frozen accident traps history to move down one path. Many agents and many options trigger an explosion of possibilities. Within lawful systems, surprises and novelty can coexist with rules.

The most dramatic example of deterministic chaos is the production of antibodies within our own immunological defense systems. For well over a century scientists puzzled about how we could generate antibodies for so many specific new kinds of infections, toxins, etc. Recently, however, immunologists have unearthed the process by which a limited number of genes can be shuffled, combined, recombined, utilizing and augmenting the power of combinations and permutations, to produce an explosion of potential outgoing trajectories. The original set of genes can be broken, spliced, reordered and rearranged almost infinitely. From limited initial conditions, unlimited creativity and newness emerge, all following scientific, biological law. Even before an infection occurs, our bodies have already prepared for it by generating astronomical combinations and permutations of antibody proteins. (This topic will be more thoroughly explored in Chapters Twenty-Seven and Sixty-One).

A more familiar example of an explosively large number of trajectories emerging out of limited initial conditions is our libraries of books, which emerge from combinations and permutations from twenty-six individually meaningless letters. In a similar manner the universe can combine a hundred or so elements to make the varied world. So it is that the immunological system in our bodies can direct the alphabet of genes to write out an Alexandrian library of varied antibodies. From limited sets of letters, elements, or genes, the cornucopia and kaleidoscope emerge. Phosphorous bearing rocks now participate in the minds of novelists in generating brand new characters and plots. Very complicated combinations of elements become thoughts in minds in living bodies. Out of a mere one hundred elements emerge tigers, palm trees, and motorcycles.

Even though it is ordered, the world is intrinsically creative and never fully predictable. The universe is not only formative, it is transformative, proliferative and inventive. (See also Chapter Seventy on Self-Emergence).

Alphabetic literacy is a product of humanity. We have seized the opportunity that nature provides, in combinations and permutations, to combine specificity with multiplicity. Our accumulated literary heritage capitalizes upon and is no more numerous than one person's cascading antibody production. Estimates of antibody production stagger the imagination. We may generate as many as a trillion antibodies every day. Similarly, we proliferate words from letters. Both literacy and antibodies demonstrate fixed antecedents, like twenty-six letters, or a limited number of genes, with almost unlimited subsequence. Everything that we will say, write or read today, all the science, poetry, blogs, and arguments, are the sound of twenty-six letters.

Our expanded concepts of lawful influence among events makes scientific thinking more flexible and more in line with daily life and common sense. We are well aware that the events in our lives are partly under control; so, accordingly, we drive safely, eat properly, chose certain neighborhoods for our homes, save our money, and get vaccinations. At the same time, we must accept implicitly the ideas that science has formalized as "top-down causality," "path dependence," and "deterministic chaos." We recognize that war and invasion can invalidate a healthy diet, that prudent savings can be erased by wild, continent-wide inflation, and that many diseases, like breast cancer, multiple sclerosis, or "ALS" kill people without our ability to predict or control the cause, despite decades of statistical research on mountains of data about millions of people. Causes of some diseases may be multiple, compounded, interactional, an explosion of possible outcomes from a set of covert precipitants, which make advanced prediction of any one case impossible.

Life seems best understood with our new ideas as an emergent expression of a creative universe that mixes law, multiplicity, probability, "more is different," "top-down causality," "path dependent history," and "deterministic chaos." Laws operate in multiple ways, and with varying degrees of precision, or looseness. Laws can also be changelings that create such new conditions that they erase their own tracks like tricksters.

Is life on Earth an improbable but also predictable end-point of routine law, as Steven Jay Gould believed; or an inevitable outcome of constrained and directed certainties, as De Duve believes; or is life on Earth something new, unique, a singularity emergent within unrepeatable, path dependent conditions? Or is life on Earth something new, yet likely also elsewhere? Are scientific laws fixed grids within which the world has its limits, or are they like solar rays streaming through the clouds for a cosmic moment? Did life on Earth elaborate new fundamental laws?

Physicist Paul Davies wrote (see also Chapter Four) in *The Goldilocks Dilemma*: "fundamentally new phenomena emerge at successive thresholds of temperature, energy, and complexity...The Big Bang is sometimes referred to as "the Creation," but nature has never ceased to be creative— spontaneous emergence of novelty and complexity guided by underlying laws." Davies is a rigorous scientist and complexity theorist who seems to be echoing the ecstatic author of the novel, *Zorba the Greek,* Nikos Kazantzakis, who wrote: "God is being built." Notice that we are discussing two things. One is the creation of new phenomena and the other is the creation of a new phenomena-creating phenomena. The maker itself may be subject to remaking.

If the "maker" is the Big Bang, how confident can we be that the Big Bang is a singular, unique, creative process? Was it a one-time thing? Will it recur? Was it a product of law, path dependence, deterministic chaos?

In *The Varieties of Religions Experience*, William James wrote that everyone asks, implicitly or explicitly, "What is the nature of the universe in which we dwell?" Now we have telescopes that can see light that has traveled towards us for fourteen billion years. What kinds of laws can possibly be guiding this light, and where did these laws come from?

Chapter Thirteen

Visions of Stridor

Just how long lasting and how far reaching is the cosmic expansion that Edwin Hubble observed? Does the Big Bang expand forever? Is there no limit to "how long" and "how far"? Does it become colder, emptier and more dispersed forever, "pointless and unfeeling?" Does the universe hit an expanding limit, and then begin to contract back down again, an accordion universe? Although Edwin Hubble's discoveries were the premiere empirical evidence for the idea of expansion, they were by no means the first time that people imagined this. Two thousand-five hundred years ago, the Buddha repeatedly spoke of an expanding and contracting universe with very long cycles. Would a rebound-contracting universe, such as the Buddha described, in the *Majjhima Nikaya* (quoted in Chapter Two), require a different set of laws to guide the developing compressions, densities, and shrinking dimensions? Wouldn't contraction necessitate different physics rules than expansion? Just how small would the world get? How hot and how compressed? Would the heat and compression again fuse electromagnetism with the weak force? Would everything then re-expand? By the same old laws, or according to newly minted ones?

The Big Bang hypothesis describes the evolution of our universe from a fiery small entity into a vast cold one, and is anchored in so much data from so many sources, that expansion per se currently has no discreditors (See Chapter Seventy-Seven, Section III). There are no scientific voices to dispute the fact that the visible universe is expanding. But the uniqueness of the Big Bang is already in dispute. Some cosmologists now believe that Big Bangs occur in an ongoing, infinite series. Paul J. Steinhardt, who holds the eminent title, Albert Einstein Professor of Science at Princeton University, and Neil Turok, who is Chairman of the Department of Mathematical Physics at Cambridge University, have written in *Endless Universe:* "The central tenet (of the Big Bang theory) that the universe emerged from a hot dense state fourteen billion years ago and has been expanding and cooling ever since, has been firmly established through many independent measurements...but the bang is not the beginning of space and time...our cosmic history consists of repeating cycles...an infinite cycle of titanic collisions...it is possible that they are endless." Interestingly, and now familiarly to us, Drs. Steinhardt and Turok, refer to themselves as boys, "...asking the same questions that have gripped people from every society, every culture, every religion, and every continent since civilization began." They also propose that some of the ordering information (Lowenstein's "initial information state"?) passes from universe to universe, each subsequent Big Bang partly formatted by its ancestor. The "bangs" are neither unique nor original.

The idea that information can be passed from one ancestral Big Bang to another contradicts the view that the Big Bang represents the emergence of space, time and information and, being the singular origin, can have no connection to any antecedent. Instead, Steinhardt and Turok have created an intricate theory in which time and space have no beginning, and some basic communication does occur from cycle to cycle. This new viewpoint by two of the twenty-first century's leading scientists, is not widely accepted, is far from proven, and requires the complicated mathematical physics of string theory. It eerily echoes the teachings of the Buddha.

Was the Buddha prescient? Two thousand five hundred years ago, he had already described something similar to our "newest" insight, that the world cycles through very long phases of contraction and expansion, and has no origin nor end, and is run by an impersonal information source of natural law. While the ancient and modern theories are not identical, they are clearly aligned.

But what is the source of natural law, inherent cosmic order, that coherent thread of information, that "Dhamma" as the Buddha called it? He encouraged his listeners to try to directly experience the universe within the microcosm of their own bodies. According to his own experience, the natural law that guides the processes of the universe also guides a code of conduct for human life. This "Dhamma," infinitely ancient, passes like crown jewels, like a constitution, from one universe to the next. We can only wonder where that eternal law was written, and on what crown those jewels first shone.

Interestingly, some theories of complex causality also existed in ancient Buddhist texts. The *Abhidhamma* is well over two thousand years old, and, in its discussion about how a person shapes his or her future Karma in accordance with actions chosen throughout life, it describes how individual choices cause complex and unpredictable karmic patterns. Since everyone makes many choices every day, and many significant choices across the lifespan, the Karmic outcome, according to the *Abhidhamma,* is a result of addition, subtraction, multiplication, and division of cause. Two choices of equal Karmic salience may augment, reduce, amplify, or erase each other. In the words of a modern translator and interpreter of *A Comprehensive Manual of Abhidhamma*, Bhikkhu Bodhi, "there is always a collection of conditions giving rise to a collection of effects."

Both complex causal thinking and cosmology have opened new windows onto wonder. The singular Big Bang, popularized by Stephen Hawking's *A Brief History of Time*, is by no means irrefutably true. We cannot assume any statement about the origins of the universe to be definitive. Concepts of endless universes, and of multiple universes compete with the notion of a unique and singular Big Bang (See Chapter Five).

Are we in the only universe, or beyond Hubble's rocketing edges, are there others, a multiverse? The idea of multiverses derives from quantum physics. Unpredictable origins may lie inside quantum domains, so within the evasive, invisible, and minute (so the argument goes) new worlds may be born.

In the micro-domain smaller than Planck's constant, matter and energy, location and motion, blur. They become probabilities instead of facts. (See Chapter Seven). There is intrinsic uncertainty. Smoothness of continuity disappears. Things happen without clear cause. Unique, discrete events happen, just happen, as discontinuities without any precursor. This was first noticed by the legendary double Nobel Laureate Marie Curie, who wondered how to predict which specific atom of a radioactive isotope would decay and emit radiation. The ultimate answer seems to be: this is a probabilistic event which cannot be predicted for an individual atom, but can be predicted very well statistically for a collection of them. We can accurately predict radioactive decay rates for a collection of uranium atoms, but the decay of a single atom remains unpredictable and without clear cause.

There is a random fluctuation, or fuzziness, in the most elementary essence of matter and energy, in the basement of the universe, as we discussed when considering uncertainty. Within the domain of quantum laws, non-caused original events can in theory occur repeatedly. Can these non-caused events include the origin of the universes? Of the theoretically possible universes that could emerge, have only one or two or a few emerged, or is it possible, as the elaborately inventive Richard Feynman thought, that over the expanse of infinite time and infinite space, all possible histories that could ever exist do in fact exist from original quantum fluctuation events that themselves have no cause? In this view, the essence of existence is spontaneous emergence out of nothingness, a process that repeats in infinite places, infinite times (see Chapter Five).

Multiverses seem more science fiction than science, more a tour de force of inventiveness than the construction of a testable hypothesis. Yet didn't the most advanced minds once believe in the reality of a geocentric little world? Could the people who argued against Copernicus' modest revolution in the relative importance of the Earth and Sun believe in the world of billions of galaxies which we now all accept as real? Might we be approaching the concept of multiverses with equally archaic skepticism? Isn't it logical that every possible arrangement of matter, energy, and law actually already exists in the universe that is a collection of universes? Is the whole scenario much bigger than we now can conceive of?

Richard Feynman's idea that all other histories are equally real was elaborated into a many-worlds cosmology by Hugh Everett, whose views were initially ridiculed and rejected. Today, the cosmology of multiple universes is championed by many distinguished scientists, like Max Tegmark at M.I.T., and Brian Greene at Columbia. Tegmark claims that, unofficially, most cosmologists support some version of multiverses.

Russian born Stanford cosmologist, Andrei Linde, has added his international reputation to the credibility of a slightly different version of multiverses, which he describes as "a huge growing fractal." Many expanding universes give rise to others which give rise to others, without beginning or end like balloons giving birth. In each of the potentially infinite universes in Dr. Linde's computations, the basic laws of particle physics and the basic mathematical constants (like the speed of light) differ. The famous Hawking view of the universe as a "singularity," a unique timespace, a Big Bang, is challenged by Linde's "eternally self-reproducing chaotic inflationary multiverses." Nobelist Steven Weinberg agrees, and wrote in "A Designer Universe?" that there may even be other physics constants in other parts of *this* universe.

Can self-important little mammals on one planet in one solar system in one region of one galaxy authoritatively and definitively rule out the possibility of multiverses?

The idea of multiverses was raised briefly in Chapter Five when we considered the logic that every possible combination of every atom can be generated in the context of infinite universes in infinite space during infinite time. If this were so, it would influence our sense of the uniqueness of our own mind. Now we are considering the theory that multiverses must indeed exist, given the infinite, causeless happenings in infinite quantum space.

On the one hand, we do not want to fixate on the comforting, comprehensible known world, as did Copernicus's wrong-minded dismissers. On the other hand, discussions of mulitverses, which are increasingly popular and backed by impressive authorities, nevertheless cannot correctly be compared to Copernicus's discoveries. Copernicus used mathematics and reason to refute Ptolemaic circles, but he used observational evidence as well. Multiverses by definition cannot be observed. Light or other physical processes cannot travel among multiverses by definition, or they would not in fact be other universes. Multiverses may provoke creative solutions to questions about why life exists, or why physical constants have the values they do (which we will return to in Chapter Forty-One), but they can never be backed up by data. Multiverses can't be confirmed or rejected, since they are beyond evidence. The concept of multiverses is mathematical metaphysics, not science.

Galaxies and cells stir wonder in me. Multiverses stir bemused incredulity. Of course, wonder and proof are not the same thing!

As far back as the nineteenth century, Henry David Thoreau prefigured Paul Davies, Nikos Kazantzakis, and Richard Feynman, and revised the old, one-shot creation myth, by writing that creation did not happen once upon a time, but is alive everywhere in every moment. Today, these words echo with the deeper meanings that emergence has been able to give them. We now believe ourselves to be in a cosmos that is *both* cataclysmically destructive and creatively self-complexifying. We cannot grasp for certain its origins or its neighbors.

Our new world, like a wool shirt, is sometimes irritating.

The idea that the universe is continuously expanding, self-complexifying, and creative, or the idea that the universe is gradually stretching from a singular Big Bang out into a cold, black, emptiness, or the sci-fi chimera of multiverses, all remove the comforts of cosmologies that have immediate sensory connection to the world. They all posit questions that we cannot answer by careful measurement or common sense logic. The

ancients imagined that what you see is what you get, that the stars that
we see are the stars that exist, that the distances our eyes probe reach out
to the horizon of the world, that the time of our life span is a meaningful
sub-unit of all time. Those old views connected to basic animal comforts.

Our modern theories are more stringent. While Copernicus reversed
the order of importance among heavenly bodies, it was Hubble's twin
discoveries that there are many other galaxies and that the universe is expanding, that shattered the primary connection we used to have between
ordinary sense experience and our cosmological intuitions. We have created visions of stridor. It is hard to believe in what we know. It is hard to
know what we should conceive the universe to be. In both the realms of
the tiny and the vast, we must acknowledge uncertainty.

How did the greatest of astronomers, (who we met in Chapter Nine),
Edwin Hubble, feel, alone at night, seated before the planet's best telescope, on a "run" of observations night after night, slowly revealing to
himself what no human being before him had ever known, that a gigantic
universe of incomprehensible spaces was still expanding into the black horizon? What was the reaction of the first man to know the galaxies beyond
the Milky Way, and to witness the razor edge of light racing into spaces
beyond comprehension? In his book, *Edwin Hubble, Mariner of the Nebulae,* historian Gale E. Christianson quotes the cutting-edge astronomer:

"The whole thing is so much bigger than I am," the pipe smoking,
former Army Major, and California celebrity wrote, "and I can't understand it so I just trust myself to it; and I forget about it."

Oh well, not everyone, particularly calibrating scientists, can express
their wonder in words and thoughts. In his solitude and stoicism, Hubble
cleaved to fact and deleted its emotional resonance. Yet, can we dare to say
that this greatest of explorers lacked a sense of wonder? Some people express their wonder by how they live, not by what they say. Who can blame
Hubble, the solitary cosmic Columbus, for his need to keep it cool? For a
long time he was the only human being who knew of the universe beyond
our galaxy. Who can begrudge this Odysseus his stiff-upper lip?

Chapter Fourteen

Exponentially More Questions

The scientific world view of the twentieth century has made our psychological universe vaster, (fourteen billion times six trillion miles wide—that is, fourteen billion light years), tinier, (the quantum domain measured by Planck's constant where Heisenberg's uncertainly principle rules), more complex, (the evolution of life from bacteria up to dinosaurs and humans), more improbable, (one hundred billion galaxies with one hundred billion stars all continuously expanding outward), more likely, (the emergence of life through the constraints that De Duve says guides chemistry inexorably towards combinatorial vitality and life), more intricate, (billions of bits of information contained in the DNA of microscopic cells), more lawful, (the four great primary forces of physics, or the widespread applicability of Einstein's general relativity), more chancy, (the emergence of weird and clanking dinosaurs on our little planet, and their extinction, probably by an improbable asteroid collision), more factual, (the vast edifice of interlocking scientific information), more incomprehensible, (the "origin" of the Big Bang, the size of the universe, the origin of life on Earth), more sewn together by interlocking, fact-based hypotheses,(the deep connections between physics, chemistry, and biology), and more fundamentally mysterious, and more embedded in our own probing mind, (the subjectivity and uncertainty, the "veil" that is present in any human thought, experiment, or measure), which, like the finger, cannot touch itself. We hover in parachutes free of beliefs and disbeliefs which only recently seemed convincing to the smartest people, but which now we are discarding and tossing out the window of our spacecraft.

We have been cut free of anchors. We are multi-oriented and disoriented.

We know so much more and have exponentially more questions. Knowledge has not diminished but expanded the realm of the unknown.

Our world-view is no longer one dimensional, or "teleological" (controlled by a belief that we personally know the goal of the universe). Stephen Jay Gould, in his "Preface" to *The Book of Life*, describes our contemporary viewpoint as: "...diverse, non-hierarchical, playful, personal, pluralistic, iconoclastic and multifarious."

Our minds are scions of a noetic web that starts with electrons and seems intrinsic to a maturing universe. Our mental powers represent the maturation of life forms, the planet, and the universe, a development that ripened over billions of years. We are children in a world that is ancient, surprising, and newly formed. Possibly science is provocative but inadequate to express the full extent of our wonder.

Chapter Fifteen

Unbridgeable Distance from the Evidence

Unlike Edwin Hubble, who was so terse, analytical, precise and scornful of imagination, there are other people who would never sit in a cold observatory's darkness, measuring galaxies for a lifetime, yet who can manage to say it all in about fifty words. The high priest of wonder is Chile's Nobel Prize winning poet, Pablo Neruda. Neruda was a citizen of the world, whose mind was steeped in Europe (he was a personal friend and student of Garcia Lorca), America (he first read Walt Whitman in translation when he was still a child and continued to read Whitman throughout his life), Sri Lanka and Myanmar in Asia (where for four years in his young adulthood he encountered Buddhism), Russia (after the Spanish Civil War, in his opposition to Fascism, Neruda became a Marxist and made many trips behind the Iron Curtain), and from his own South American native peoples, who had their own advanced astronomical observatories. He wrote (in the collection and translation, *Neruda, Selected Poems*, edited by Nathaniel Tarn):

> I sought you, my father,
> young warrior of copper and darkness...

> Then up the ladder of the earth I climbed
> through the barbed, jungles' thickets
> until I reached you, Macchu Picchu.

Although the specific history and functional role of the ruins at Macchu Picchu remains debatable, the voluptuous fortress in the Peruvian Andes was, among other things, an observatory, in which many architectural alignments speak to an advanced knowledge of astronomy among pre-Columbian Americans. Neruda placed himself in lineage with these pagan scientists of the sky.

Neruda wrote that he witnessed the unfastening of the heavens and beheld the universe shot through with arrows, fire, and flowers. Although he felt himself to be only an "infinitesimal being," he described, in his signature poem, "Poetry" (as translated from the Spanish by Alastair Reid) that he became intoxicated, envisioning the star-studded void, and:

...felt myself to be a pure part
of the abyss,
I wheeled with the stars,
my heart broke loose on the wind.

Unlike Hubble's noncommittal literalism about the implication of his discoveries, Pablo Neruda, like his hero, Walt Whitman, felt that cosmic imagery expanded the stage upon which the human drama unfolds. Free of the dismay and existential nausea of European civilization, Neruda, self-proclaimed son of copper and darkness, could find in the new dimensions of twentieth century science an ecstatic release from boundaries and confinement. Rather than disorientation, he found *fusion* with the universe and *release* from the confines of his personal self. He poeticized Einstein's "cosmic religious feeling."

Our literal measurements of time and space keep expanding and with them our poetic stage. Science may have expanded our minds into realms that defy mere reason and that demand more personal answers.

Hubble augmented our sense of the universe one billion-fold. Einstein revealed the world as an entity, an egg containing space and time, rather than a collection of units riding in a pre-existing space and time. Eventually, Maarten Schmidt, using the same red-shift observations as Hubble (See Chapter Nine), but with the improved telescope on Mt. Palomar, in 1963, discovered quasars, whose exact nature still remains controversial. To our Earth-based instruments, that use visible light or radio waves, quasars appear dim, but that is only because they are extremely distant, located far beyond where Edwin Hubble had thought the universe ended. Due to their large red shift, Maarten Schmidt realized that, to be visible to us at all, these far-off objects had to be very bright, very old, and very far away. Today quasars are thought to be ancient galaxies streaming into their core black holes, emitting vast amounts of radio waves, and they can be as bright as a *trillion* suns.

In our new worldview, the expansive scriptural poetry of the *Bhagavad Gita*, with its light of a million suns (Chapter Two), now reads as a minimalist understatement. Quasars emit bright lights from distances, which are located many billions of light years away (that is, billions times trillions of miles). Because their light has taken so long to reach us (many billions of years), when we observe quasars we are seeing cosmic events that actually happened billions of years ago! One of the most difficult aspects of astronomy to actually comprehend emotionally is that, because instruments like radio telescopes or the orbiting Hubble telescope see so far away, and the light by which we see has traveled so long, we are actually seeing the past. Maarten Schmidt saw quasars, galaxies that may have actually imploded and disappeared billions of years ago, possibly even before our sun was born!

More recently, probes of light, space and time using technologies that do not rest on Earth nor on visible light waves, such as COBE, the cosmic background explorer, or WMAP, which studied cosmic microwave background, from the platforms of satellites, have informed us that the universe is approximately thirteen and seven tenths billion years old, and approximately equally thirteen billion light years large. Some scientists who accept the *age* of the universe as approximately fourteen billion years believe that the *size* is much greater, as much as forty-five billion light years in radius, distances we can never see or contact in any known way, since even light (or other electromagnetism) would have to travel for forty-five billion years to get there! (One light year is approximately six trillion miles). The universe may be too young for light to have crossed its expanse. The cosmic microwave background radiation that we can actually observe by WMAP is as old as the Big Bang, and is in fact the key proof that a Big Bang occurred and scattered electromagnetic radiation somewhat uniformly throughout the entire contactable universe. The universe is filled with a small amount of radiation heat left over from its explosive stage.

Though we can bandy words and numbers around, we cannot hold in mind or give meaning to the universe that science has revealed around us. We routinely see the past. We see things that may not exist anymore, in fact, probably do not exist. We see further than we can give intuitive comprehension.

Do you relate differently to Hubble's universe of one billion light years, that is, one billion times six trillion miles wide, than to Schmidt's universe which is twelve to fourteen billion light years wide? Carl Sagan tried to dispense to television viewers a world of billions of galaxies, each with billions of stars. It was these types of numbers that empowered De Duve to speak of trillions of planets and ecosystems. Into this trans-conceivable new vision of the universe have been added black holes, worm holes, dark matter, dark energy, and anti-matter, which *defy our experiences*, and are premised on *logical deductions*, *mathematical relationships*, and *empirical evidence from experimental tools* we never see and whose proper use we cannot validate, like infrared telescopes. (In Section VI we will discuss possible roles for dark matter and dark energy).

Rigorous, empirical, fact-based cosmology becomes increasingly puzzling and elusive to our own lifetime-time-span, or gravity-grounded world. Our knowledge is increasingly divorced from our immediate personal experience.

Not only the content of science but its manner of creating new information is also generating unslakeable poetic wonder.

Science is supposedly based on evidence of the senses. In this way it differs from imagining, but also from reasoning. The refutation of a hypothesis is supposed to be accomplished through testable, confirmable data acquisition. Hundreds of years ago, this testability rested upon experimental results that anyone could replicate. Look through Galileo's telescope and you, too, can see the moons of Jupiter. Even Einstein's theory of General Relativity was tested by relatively simple data. Look at Sir Arthur Stanley Eddington's photographic plates from 1919, and you, too, can see evidence that, during a solar eclipse, when the sun's rays were momentarily darkened, light rays from stars passing close to the sun's great gravitational field, were bent, portraying background stars in "displaced" positions, adding firm evidence to Einstein's math-based deduction that gravity bends light.

Today, however, most science rests on sensory amplification by instruments that even their users cannot construct or fully comprehend or vouch for. Microbiologists who use electron microscopes are not electronic engineers who can guarantee the validity of their tool. Cosmologists who use the data sent back to us by the Hubble space telescope are not in the same team of people who design, build, and launch an orbiting satellite. Furthermore, data sent back from modern telescopes is too complicated for any one person to assess. At NASA's advanced supercomputing center, 128 computer monitors create an integrated hyperwall of data. Even when sciences continue to utilize tangible, visible data—let's say fossil bones—the deep development of scientific disciplines requires years of study, PhD and postdoctoral specialization, to integrate the evidence with mountains of data and interpretations acquired over decades and centuries, and which require years of study to understand. Not only the data, but the tools by which it is gathered and the concepts that invest its meanings, evade our quotidian experiences.

Our scientific world-view, which originally freed us from argument based on belief and authority, has again made us dependent for our knowledge upon a priesthood of sorts. We once again are forced into a position where there is a fundamental question about verification and trust. We no longer have direct access to the very "facts" upon which we hope to accurately construct our psychic world. This unfortunately has led to a backlash of unabashed, anti-scientific fantasy, justified as faith-based belief, meaning capricious fantasy without any recourse to evidence at all.

In his biography of Einstein, Walter Isaacson captures in humorous anecdotes this transformation of scientific evidence. In 1931, when Einstein visited Edwin Hubble at the Mt. Wilson observatory in California, Elsa Einstein accompanied her husband, and was told that the observatory's massive technical panoply of telescopes, levers, lifts, and dials were used "…to determine the scope and shape of the universe." She reportedly replied, "Well, my husband does that on the back of an old envelope." This anecdote is not just a joke at the expense of Mrs. Einstein. It is a measure of the change of the locus and attributes of truth-seeking behavior. Einstein used mathematical reasoning. Hubble used technological amplification of electro-magnetic waves in the spectrum of visible light.

When the great Einstein finally fled the darkness of Europe's pending genocide, to take up residence at The Institute for Advanced Study, at Princeton in 1933, he was asked to submit his equipment requisition. "A desk, or table, a chair, paper and pencils…oh yes, and a large wastebasket…" Compare that legendary elegance and simplicity to the thousands of people and millions of dollars behind WMAP, launched in 2001, the microwave portrait of the universe, which observed the universe billions of light years away, and therefore billions of years ago, from an orbiting satellite with a bank of detectors and computers and teams of scientists working for years to tell us (who can never reassess the evidence) what it means. Our contemporary information base is empirical, subject to test and challenge, but it is also based, for most of us, upon extrapolated *trust*. We do not have direct access to, or the ability to assess the information that forms the basis of our worldview. Exploration and authentication have passed out of the domain of the individual into the collective and large scale. For better and for worse, this impacts our conviction, which is both weighted down by authority, but also decreased by our unbridgeable distance from the evidence.

Chapter Sixteen

It Takes a Civilization to See a Galaxy

hen we see photographs of other galaxies on the front page of *The New York Times* taken by the orbiting Hubble telescope, (named of course in honor of Edwin Hubble), or when we watch an IMAX 3D movie about the Hubble's discoveries, we are privy to images that took tens of thousands of people decades to create. No one, no group, can launch a telescope to rotate for years three hundred and fifty miles above the Earth while photographing the universe thousands of times. Such a feat requires the aggregated scientific knowledge of our civilization, the wealth of America, the legislative funding processes of the United States Congress, the administrative organization of the National Aeronautics and Space Agency, as well as a city-large collection of workers, technicians and scientists. Our visual field was clarified and expanded by the orbiting Hubble telescope, which rests upon a civilization, which like a fungus, has a tiny mushroom top and miles of mycelium underground.

Each Hubble photo of a spiral galaxy is the product of long aggregated and long entwined minds, technologies, finances, and leaders. The photo is a triumph of a collective enterprise with myriad roots ascending up to one focus.

The webs of galaxy clusters spanning hundreds of billions of trillions of miles that the Hubble has clearly pictured, provide to the common man a portrait of a reality completely inconceivable even to an Einstein a mere eighty years ago. The viewer is edified, and promoted beyond yesterday's greatest sages, and yet the viewer is also demoted to a stunned and helpless dependency upon an information-gleaning process that escapes from one person's power to hold.

No wonder that so much of humanity simply denies, turns away and ignores the doubts, implications and cultural revolution necessitated by the absorption of such visions of the universe. To integrate the visually vivid cosmology of the essentially endless web of essentially numberless stars—the face of infinity and eternity—we have to not only radically transform our image of ourselves, our historical moment, our civilization, our planet and our religion. We also have to pay obeisance to the illimitable network of taxpayers, legislators, high school science teachers, computer geeks, physics professors, lens makers, and these, too, are only the enumerable groups for the pyramid of skill and knowledge from the tip of which the photo is snapped.

What we think we know is embedded in a matrix mind. We know more because we are no longer solitary watchers of the sky. The Hubble Telescope is the eye of an era. It sees with a retina that is spread out across a civilization. Every high school math teacher who once instructed a NASA scientist is one of the pixels on the big screen of the cosmic portrait which is imaged, transmitted, downloaded and processed as Hubble-derived imagery enters one single viewer's eyes. It takes an era to see the galaxies. Even in the 1920's, Edwin Hubble on Mt. Wilson stood on the shoulders of other astronomers, lens grinders, and telescope makers. No one can see a galaxy alone. All other galaxies are either hazy or invisible to our own eyes. It takes a civilization to see a galaxy.

Under this modern condition, our sense of wonder plunges and rises like a dolphin. We try to believe in things we cannot understand and for which we have no direct evidence, because we believe someone else understands them and has evidence for them. We have uncovered a Pandora's box of data, ideas, images, facts, hypotheses, all of which lie beyond our personal power to affirm or refute. When we ask ourselves what we know, or what we think we know, are we standing within a scientific world-view, or are we dependent, faith-based believers, in a cathedral of science whose architects are invisible to us? Have we become bold individualists free of ecclesiastical dogma and authority? Or are we sheepish followers of a new esoteric priesthood?

The cosmos may be essentially empty and obliterative, a place of "chaotic violence (Sagan)" on astronomical scales. Neruda can wheel with it. But when Sagan describes, "In every quasar explosion, millions of worlds, some with life and the intelligence to understand what is happening—may be utterly destroyed," we might find ourselves wheeling-challenged.

We have replaced the archaic, superstitious, and fanciful folklore of old with a rational, experimental, evidentiary incomprehensibility. We cannot wrap around with meaning the package that we claim to understand. The zones of conceptual thinking, evidence, assumption and understanding are being tugged apart rather than drawn together by our expanding scientific enterprise.

We now have a long enough history of errors that have been corrected, like the geocentric universe, or like the single-galaxy universe, to recognize that we can easily once again fall prey to our own ignorance. Just as former beliefs were founded on limited information and culturally formatted assumptions, our current world-views may be also. All thoughts, all sentences, all knowledge is framed in the context of culturally embedded narratives, while at the same time, not all narratives are equal, differing in the amount of evidence they assemble, the manner in which the evidence was acquired, the feasibility of checking and correcting the evidence, the mutual integration of disparate sub-components of the narrative, the rigor of the logic, the degree of self-reflective and publicly assessable disclosure of bias or sub textual motive on the part of the investigators, and the credibility and interconnectivity of the new narrative with other antecedent and overlapping statements. How can we assess all of this when we are told that we are looking at a photo from the Hubble Space Telescope that pictures an unthinkably distant spiral galaxy of billions of suns? How can we affirm or refute the legendary Richard Feynman's imaginative logic about quantumly acausal multiverses? How can we accept or reject Turok and Steinhardt's mathematically arcane string theory from which they derive their argument for an endless series of Big Bangs? (Chapter Thirteen).

Yet at the same time, our integrity, and our sense of wonder, will not allow us to collapse into infantile father fantasy, nor cynical exhaustion. We have also been exposed to a world pregnant with invention and wonder. We are no longer soothed by informatically impoverished fairy tales without evidence, whose narrative bluntly contradicts an accreting mountain of data. We do not care to live developmentally arrested by adhering to folklore. Neither are we satisfied by the helpless sigh of agnosticism.

Our sense of wonder is not limited to physics, astronomy, and cosmology. Biology, the study of life, exploded in the twentieth century, and revealed a universe inside of us. The complexity of life that we have uncovered is as staggering as the cosmos. (This plunge into the wonder of life will be the focus of Sections III, IV, and V).

The expansion of our sensorium, through technology, has been as stunning on the small scale as on the large scale. We now study life, as we study the cosmos, with information-amplifying techniques that we cannot understand or validate, but which we also can't ignore.

Life holds *order, unity, creativity, resilience, duration*, and *slow elaboration of awareness*. Out of nothing: something that does not seem out of nothing. Life on Earth, our life, feels compellingly valuable and monumentally compounded. One oak leaf is a tinker toy that contains more bits of properly aligned information than our minds can grasp. There is a universe of information necessary to activate the leaf. The forests of New England, or of anywhere else, are themselves like galaxies in which the numbers of properly covalently bonded atoms are far more numerous than stars. Then come the complete, functioning leaves, breathing out the O_2 of life. Life now appears as momentous and complex as the expanding cosmos itself. The dimensions and qualities of the cosmos stun us into wonder, and so do the minutiae and details about the intricacies of the cell, our cells. Our new world-view provokes us to think large and small, about a world that seems profligate with galaxies and pecunious with molecules, none of which we can directly touch or see by ourselves.

We stand between a void and a biosphere, emptiness and vitality, frozen eternities and the hum of cities. We are born out of a macrocosmos and a microcosmos with which we are partly familiar but which remain partly veiled. Meaninglessly vast spaces seem to have given birth to the minds of Neruda, Whitman, Pauli, you and me. We find ourselves orienteering between supernovae explosions and nursing mothers. Is Walt Whitman's "robust soul" formed in concert with, or made hash of, by the quasar-far eternities? Theologians say it all proves god. Material reductionists say its all a quirk, a lucky jumble of pick-up-sticks. Can you believe that either of them actually knows?

How can we hold in one picture the emptiness of trillion-miled space, with the fullness of our sense of meaning and love? Even a backyard rabbit seems to contradict the cold, pointless, expansion of the universe that we also take as real.

We and the rabbit are *compacted creations* that required every apparently sterile law, from electromagnetism to Pauli exclusion, and from billions of years of development of cells. How can so much order, and compression of order occur in a dissipating explosive universe?

Our search for understanding, whether based on poetic, intuitive methods, or upon scientific, experimental methods, is all subject to the limitations of our main tool: the human mind. Rather than trying to prove or disprove any physics or biology, this Section I is focused on a psychological point. Our understanding of the world is historical, contextual, and information-based, all of which are constantly changing. This modern constructionist sensibility alone casts into doubt, or cautions against, rigid beliefs and conclusions, but there is also the compounding factor that the cosmos itself may emerge with state-specific phenomena, and may be self-complexifying. *The evolving world may not contain a "final knowledge," because it may not be fixed into any final form.* The mind by which we hold the world in wonder is neither necessarily objective nor final. There is no final, mother Einsteinian law because the initial information state has already been erased and transformed by top down causality and frozen accidents. Our wonder about the world may be quintessentially suspended. Like cottonwood seeds, we float in wonder.

Chapter Seventeen

This Much We Can Say

It has taken our planet four and a half billion years or so, to construct the human mind, or to construct the human mind's capacities to download mind out of its cosmic matrix. Yet our planet, sun, and solar system are "new," second-generation (or multi-generation) cosmic citizens whose ancestor stars blew up and created the pool of materials out of which our own platform-planet was formed. There already exist elsewhere many stars and planets that have been around much longer, and that have had much more opportunity to evolve and to have beings who "know."

What does an extra billion, or four billion years of evolution add to the capacity for insight into the cosmos? What might those ancient beings, in other, older galaxies, if they exist, know? Do they have Einsteins who can solve once and for all the equation for a single unified field theory that explains the entire world in one formula? Do they have Nerudas who can poetically explain how to dance amidst the unimaginable destruction implicit in black holes and quasars?

Similarly, if life on Earth should avoid global warming, nuclear war, overpopulation, resource depletion, and environmental pollution, and evolve for another ten thousand years, or a million years, or a billion years, what will we, or should I say, "they," know? Will it still be true that, as d'Espagnat has written, "Mystery is a constituent property of knowing"? Will the self-complexifying, creative cosmos of Philip Anderson, Harold Morowitz, and Paul Davies stay ahead of and continue to mystify the branching-brained hoary sages? Or will those wise and by then very ancient beings know fully well what Einstein called "the mind of God", or what Buddha called "the Dhamma"? Will they have tragically accepted Stephen Weinberg's "pointlessness", and Jacques Monod's "aloneness in unfeeling immensity"? Or will they wither into internecine violence, fear and stupidity, like the fictional denizens of H.G. Well's novel, *The Time*

Machine, or of Isaac Asimov's "Nightfall"? Will their wonder collapse into facts and cynicism, or will it exfoliate into continuously expanding horizons?

We hover among subject-object uncertainty, and limitations to our essential capacities. Our knowledge may always be separated from ultimate reality and truth by the veil of our bio-cellular neural mechanism, the brain. Do we know, or do we only know what we have the capacity to know, which may be far less than, or equal to, the knowable?

Despite our plight of trying to stand upright on an ocean, to surf waves where there is no shore, we have also discovered a great deal. If we do not stand on solid ground, then we have learned a lot about our native seas. *This much we can say*: Whether we build up our understanding of the world from the physical laws that guide its operations, or from the quantum physical laws that shepherd the probabilities of its matter-energy matrix, or from the organization of the world's building blocks into the atoms, molecules, and cells of life; whether we are trying to understand our own embodied lives, or the expanding galactic immensities, we encounter an incalculable compounding of order, an intricate historical lineage, and the shining forth of awareness. *Complexity*, *time*, and *mind* are our great discoveries.

We are light years away from randomness. A combinatorial explosion of lawful interactions among transcalculable atoms over billions of years have created a vibrant planet of life. We are combinations, laws, and time self-aware.

We see the indelible signature of order within the frozen and exploding universe.

Our minds are as curiously alight as quasars. The architecture of existence is built on micro-galaxies within each one of our cells. For all of our growth in positive knowledge, we do not yet understand Whitman's leaf of grass, or Morowitz's "mind." Every cell with its vital humming remains inexplicable to us. We are the products of vaster dimensions within us than there are around us. We are magnificent with information. Within our DNA are canonical collections of evolution's long savored wisdom.

Therefore, wonder is not merely a pleasant mental state, but a fundamental orientation and realization. We relish our encounter with the world we are and do not know.

Wonder moves across our mind, recognized and unrecognized, with the energy supplied by "ATP," which is the coin for wealth pouring into us from the sun, our energy source, which got its energy-source hydrogen, from supernova explosions of previous suns, which got its own hydrogen from where, we wonder. That we are stirred by wonder also begs an explanation. What, psychologically, is wonder? How has it come to be that we can hold, in our mind's eye, the world in wonder?

SECTION II

What is Wonder? The Psychology of Wonder

Chapter Eighteen

A Deep Echo

A man and woman from North America are standing on a high plateau, surrounded by rugged mountains in South Africa. The mountains are formed of sandstone that has been on the surface of the Earth for a very long time. Throughout the approximately five hundred million years during which the sandstone has been exposed to the elements, it has been carved into pinnacles, valleys, and blocky mountains. South Africa was once the center of an ancient super- continent, called Gondwana, and its central position within that bigger landmass left it relatively stable, as plate tectonics split the paleocontinents apart and formed new patterns of land and ocean out of the movable pieces of the Earth's crust. The geologic conservatism of the southern region of Africa led to geomorphology that is free from the smashing, bashing, and uplifts of the Appalachians or the Himalayas. The mountains of South Africa are the gnarled residue of half-a-billion years of continuous rain, wind, cold, heat and ultraviolet light, producing a primal moonscape of elemental antiquity. The mountains are dry, unforested, exposed and jagged. Here it is easy to feel that the Earth is an old rock hurtling through space. Now it is night in Southern Africa. The Americans are standing with their heads craned back, looking at the stars.

In their six decades of life, they have never seen these stars, nor any stars like these. The altitude, the dry air, the distance from any major city, and the viewpoint of the deep Southern Hemisphere, where a person can view the brightest part of the Milky Way's disk nearly straight overhead, reveals a new sky to them. It is spangled with stars; it is awash with stars. Only vaguely given orientation by the constellation of the Southern Cross, which can be seen only in the Southern hemisphere, the sky is punctured, felted, fur-coated with stars. There are many stars, more stars (it seems) than any other night or any other sky, big jeweled diamond stars, and sequined washes of winking stars, regions, zones, and lakes of stars, milky nebular stars pin-pointing the interstices between bigger stars, tiny stars disappearing into the eye's inability to resolve them, vanishing pin points kindling the sky. The sky is both black and swallowing, and alight and signaling, an impossibly alien sky, a beckoning semaphore sky that twinkles in friendly incandescence, a starship sky, a sky from which every point of light shines into the brain and lasers away thought or comprehension. It is Ralph Waldo Emerson's "believing and adoring" sky, Rachel Carson's "misty-river-of-the Milky Way" sky, Edwin Hubble's expanding-light-years sky, Carl Sagan's inhabited sky, Isaac Asimov's terrifying sky.

Unstabilized by their backward arched necks, the teetering Americans can keep themselves aloft only by letting their minds splash into a rippling ocean of wonder. Wonder halos them with its phosphorescence. They imagine themselves to be ancient Greeks, among deities who are constellations. They feel themselves to be deities crowned with the halos of heaven. They are tugged into becoming wispy and fading ghosts blown away into the void. Wonder pulses like ripples from a stone. They are standing at the shoreline of ultimate mystery and knowledge. What should they do? How must they now live in consequence of this revelation?

"I'm tired. It's cold. Let's go in."

What is wonder? Why do human beings contain this emotion, or thought-pattern, or absence of thought pattern? Of what use is it? Presuming that people are mammals, products of evolution and adaptation, what survival utility does wonder contain? We can easily see why certain human emotions, like fear or love, are biological tools, calculated to the avoidance of danger or to the establishment of safety-facilitating, cooperative families and groups. Why do we wonder at the sky, and then shut down, exhausted, after a period of over-indulgence, as if these glories and cessations of wonder were also evolutionary adaptations?

Is wonder non-Darwinian, not an adaptation at all, but an echo from a different zone of truths? Is wonder the same as, or different from, amazement, awe, curiosity, or delight? Is it a mere pleasure, or is it a reliable directive, providing useful information in the same way as sight and hearing do? Is it merely subjective, aesthetic, fanciful and cultural, or does it contain any enduring, objective truth-value about the outside realities around us? Why do we cross latitudes and hemispheres to indulge in moments of wonder, then shut it down, turn it off, only afterwards to treasure its memory for years?

Part of the definition of wonder is its power. When we hear that some-one has had an experience of wonder, we expect it to remain important to them. If we hear that our friends took a vacation that filled them with wonder, we expect to see their photos, and to hear about ongoing follow-up reading, or study; or we expect our friends to return annually, or at least recurrently, to the site where wonder descended into their midst. When we hear that a college student found a topic that filled her with wonder, we expect her to pursue that study, maybe to major in it, possibly to follow it into a career. We intuitively separate wonder from its weaker and more intellectual cousin, curiosity, which we call "mere curiosity." Curiosity killed the cat, but we do not attribute wonder to domestic pets. Wonder, as we use the word, is explosive and vectoral. It has psychic pop and endur-ing direction. It beckons. We seek wonder, return to it, and follow it, even though we also become depleted by it and need to shut it off.

Notice how wonder differs from its cousins, astonishment and amazement. Astonishment refers to being taken by surprise in a relatively positive way when something new breaks into our psychological world. Astonishment does not connote a powerful and enduring impact. Quo-tidian and transient surprises can astonish us, like the time a barred owl suddenly flew down out of an apple tree in front of my window in mid-day. Astonishment refers to surprises of containable magnitude. We are surprised, but not required to *re-orient* our beliefs or our lives.

Amazement is a term that is generally used when the surprise exceeds astonishment and induces temporary disorientation, disjunction from pre-vious assumptions, a gap in comprehension. If we fully understand what is happening, no matter how unexpected, we are not said to be in amaze-ment. I was, in fact, amazed to learn that albatrosses can fly for months at a time without ever landing on solid ground to rest. Unlike astonish-ment at the appearance of an owl at mid-day, this fact about albatrosses required me to shed my beliefs about muscle strength, sleep, and rest. How could any animal stay suspended for such an unbelievable period of time? Amazement edges towards wonder. Still, amazement is not long lasting. I subsequently found out that albatrosses can lock their wings in an outright position, using only bones, and so stay aloft without using their muscles, so that they actually sleep while flying on cruise control. Generally, by amazement, we mean we have been pleasantly surprised in a way that stays for a while and then passes away. Our amazement self-reduces either by depletion of the novelty, or by new information and re-organization of our perceptions. Most of us enjoy the amazement contained in books about oceans or in television shows about outer space. They are educative and thought provoking in the evening, and in the morning we can return to our chores and work without feeling bothered by them. We do not have to delve into their subtleties or feel obliged to answer to them.

The lovely cousins, astonishment and amazement, have become ritualized in contemporary spiritual practices. Recognizing correctly that routine beliefs and perceptions confine us and make us stale, some New Age psychology has emphasized the cleansing value of fresh perceptions, disbelief, and a mind free of thought. In these teachings, a blank mind is assumed to be a receptive and renewable one, and an absence of thought, with its subsequent return to freshness of perception, is seen as a form of revelation. Certainly, it is important to relinquish over-rehearsed and overvalued dogmas, whether they are intellectual or perceptual, and to encounter the world in unmediated directness. This exercise of refreshing precognitive directness is a welcome part of openness and the ability to learn. But a life enamored of the "A-twins," astonishment and amazement, makes a wide angle away from wonder, which even at the outset, we have agreed to define as salient and directive.

Wonder is the word we preserve to refer to events that provoke a deep echo, that make us tremble. Wonder is a signpost at a crossroads. Due to the experience of wonder, we change direction. Our term refers to experiences that forge the subsequent days and years, branding them with memory, reverence, and, yes, with intelligent, thoughtful intellectual pursuit. Far from making your mind go blank, wonder refers to experiences that catalyze urgent, ongoing consideration.

Wonder has no important connection to "mindfulness," which has entered the English language recently and means paying attention to what you are doing. Before the "mindfulness" trend, people had long understood the importance of concentrated attention. Re-labeling simple alertness and calm attention with Buddhist-flavored terms like "mindfulness" has gained cachet in a society riven by multi-tasking and the intrusive distractions of cell phones and emails. Congratulations to those people who can establish mindfulness in the slurry of contemporary static. Such an attainment may well be a precursor to moments of wonder, but a precursor only.

Wonder is not the same thing as being delighted by the passing magic of a graceful movement, no matter how charmed. Nor does wonder mean "being in the now," because we need a term, like wonder, to help us refer to experiences, feelings, and thoughts that roll powerful tsunamis through our psyches. Even stargazing generally is little more than transient entertainment, a form of nocturnal jewelry. Wonder is our referent for experiences with staying power. It is not only "in the moment," but it rides the crests of many moment's ongoing waveforms. Wonder refers to experiences that *endure* in your psyche, that hitch a ride and won't let go. Wonder gives momentum to a string of serially surging "nows."

In our attempt to separate out and clarify wonder from words that refer to its approximations and dilutions, we have at last come to its trickiest relative. Wonder is often paired with awe, as in the expression, "awe and wonder." They are more than cousins: Siamese words, maybe, with some inseparable isthmus of tissue between them. They are casually bandied about as synonyms. But if we want to stand in wonder, we would be wise to give awe a shove, and if we can't push it behind us, at least try to give it some distance, because awe and wonder have different flavors. By "awe" we generally mean something tinged with fear or danger, as well as with wonder. Awe is sometimes called, "solemn wonder." We are not comfortable with and delighted by awe, although we may simultaneously be renewed and fundamentally changed by it.

When the god, Krishna, revealed to Arjuna, in the *Bhagavad Gita* (Chapter Two), a vision of cosmic unity, in which everything that moves and moves not is part of a single whole, but which by extension includes all-devouring time, which swallows all living things like moths rushing to a flame, Arjuna's state "hrstaroma," or "hair standing on end" is awe. But maybe it is also "awe and wonder." Arjuna, mortal like us, has his hair standing on end like an alley cat who sees a rottweiler, but he also becomes inspired, relieved, a knower-of-liberating-truth. Maybe this is awe-followed-by-wonder, or rapid oscillation of awe-then-wonder.

In contrast, our North American star gazers in South Africa are gasping in wonder, but a little edge of awe is moving in and out of their consciousness, a serrated, eerie intrusion, that may well be partly responsible for the exhaustion, and the need to bang down a lid across heaven's revelations.

If we cannot neatly compartmentalize awe from wonder, then at least we can try to tilt the board, so that our marbles are more likely to roll to where we want them. Pure awe stands, holding right hands with wonder, but left hands with dread, "awe and dread." Wonder and dread have awe conciliating between them, but are on a continuum that is hard to neatly slice. Still, if dread arrives, wonder has already departed unnoticed.

Does this help us understand Edwin Hubble, cosmology's greatest revolutionary, whose line of sight was the first ever to exit from our galaxy and to enter the cosmos, and who removed the chains of stasis from our perception of the stars to reveal their mystic rocketry? Both in his writing and in his daily life, Hubble was among the most confined and prosaic scientific factualists (Chapter Thirteen). Given his astronomical courage and isolation, it may well have been necessary for him to psychologically drag the Earth with extra-heavy anchors. Had he allowed into his mind any further deviation from the beliefs of his contemporaries, might he have been impelled to flee with awe and dread from his discoveries and revelations? Don't we all contain a wonder dial that can rotate up into a red–zone of awe and dread? Doesn't the expansive mood of wonder nevertheless require limits? What was Edwin Hubble supposed to think or feel with his eyes and his mind a billion light years away, beyond those of any other human, beyond the skies that gods once rode, beyond even the dreams of theology?

In any case, we have finally encircled wonder, dusting aside competitors, imposters, and family members, but not yet plunging into its direct and definitively powerful vortex. We want to understand wonder, the *forceful, welcomed, enduring transformer*, the badge of our best moments.

Let's return to our North American stargazers in Africa and figure out, psychologically, what they are doing as they stand there under a sky of wonder.

They have placed themselves somewhere new. The history and landforms are astonishing and amazing. They have been pleasantly unanchored, as well-cared-for travelers often are. They are, as they want to be, learning something. But the Southern night takes them by surprise. The containable mental states of astonishment and amazement are inadequate to hold the feelings pouring into them from the zone of the Southern Cross. Incomprehension sets in:

Can this really be true? Are there so many stars? Why have I never seen them before? Is the sky so different here? Or am I just seeing more clearly? What can it mean, all those stars? Can they really be light years away, hundreds or thousands or millions or billions of light years away? Is this why ancient people were religious? Am I religious? How can there possibly be so many stars, planets, galaxies, worlds? Why make them all? How did the world get here? Can the Big Bang really be true? Will the origin of the universe be explained differently some day by a new Newton, Hubble, Hawking, Einstein? Even if the Big Bang is true, how did it get here? Why did it happen? My god, this is beautiful! Look at those subtle colors, the twinkling metallic ruby and silver in the little points of light. I can't see them. I see them and can't see them. I'm not sure what I am seeing. I can see that I am seeing at the edge of my sight. Whatever I focus on, limits my ability to focus on its surroundings. Whenever I change my point of focus, I clarify one thing and blur another. Too bad I don't have a telescope. But then the same thing would happen anyway, other small lights would move back beyond resolving power. Besides, it's the panorama, not the detail. Look at that sweep of star-spangled space. But no, it's the details within the panorama. Look how many stars there are in each small region of space. Can I really be alive for sixty years in a world that is billions of light years across and billions of years old? My god. My god. I can't think about this. But I have to. My god. Look at those stars!

Out of a powerful, beautiful, unexpected, unknown array of stimuli, disrupting perceptions arise. Some pre-existing psychological gestalt shatters. Comfortable ideas, familiar as old shoes, tear. The mind moves back and forth between perception, emotion, cognition, and a-cognition, seeking to receive, to hold, to be aware, and to understand. Ecstatic emotions expand out to their limit, then surge and fade in a series that runs towards depletion.

The man and woman in wonder struggle to retain their revelatory connection to the flood of beauty and incomprehension. Sometimes they are receptive and clueless at the threshold of the universe, and sometimes they must thrash through everything they know, to put comprehension back into their mind's nexus with the universe. They are in love, and exhausted.

Wonder is not limited to starry skies and natural scenes.

Chapter Nineteen

Shuddering at the Sight

Wonder as we have seen, is classically associated with childhood, and with stars. The two often fuse, as in "Twinkle, twinkle little star, how I wonder…" But wonder can also be adult and intellectual. Look at reading.

The 19th century English romantic poet, John Keats, had often heard about the poetry of "deep brow'd Homer," but was unable to appreciate the poet in either his original Greek, or in English translation, until Keats came upon translations of the *Iliad* and *Odyssey* by Shakespeare's contemporary, George Chapman. When Keats heard Chapman's text "speak out loud and bold:"

> Then I felt like some watcher of the skies
> When a new planet swims into his ken;
> Or like stout Cortez when with eagle eyes
> He star'd at the Pacific—and all his men
> Look'd at each other with a wild surmise—
> Silent, upon a peak in Darien.

Despite his historical error, placing the genocidaire, Cortez, where in fact Balboa belongs, as the leader of the Spanish expedition of 1513 that crossed the Darien isthmus of Panama and became the first Europeans to reach the Pacific overland, Keats' poem has been canonized as an expression of wonder that wafted up from literature. Still, it is notable that Keats needed a nature image of muscular discovery to convey an experience he had while sitting in a chair and reading. Starting once again with the heavens, Keats abandons them and coins his echoing description of wonder:

> "…with wild surmise—
> Silent upon a peak in Darien."

The young and effusive romantic poet, Keats, linked his sense of wonder to his concept of "negative capability." Negative capability is a capacity for accepting uncertainty and the unresolved, "which Shakespeare possessed so enormously." Keats further defined negative capability as, "when a man is capable of being in uncertainties, Mysteries, and doubts without any irritable reaching after facts and reason."

Wonder shares with Keats' concept of negative capability the holding capacity, but wonder does not exclude fact or reason. While Keats correctly observed that great poetry derives from accepting uncertainty, he rested his sense of holy authority on imagination. He used a metaphor, "A Mansion Of Many Apartments," to express the multiple potentials within the fully understanding mind. Keats' insights have persisted into the twentieth century, influencing philosophers and educators like John Dewey, who promoted, "intentional open-mindedness." However, Keats enshrined his own deity: "Beauty overcomes every other consideration."

Keats certainly had a sense of wonder and could express it memorably, but his concept of negative capability, and his romantic poetic philosophy are not exactly the same thing. Wonder is similar to, but not the same as Keats's attempt to define it as beauty coupled to "negative capability."

Intellectual wonder is not rare. Take the case of Subrahmanyan Chandrasekhar.

"Chandra" was, like Edwin Hubble and Carl Sagan, another University of Chicago contribution to astrophysics. At the age of nineteen, on the boat on his way from India to start his PhD at Cambridge, the boy formulated the thoughts which have become a landmark in cosmology, and which are known as "the Chandrasekhar limit." He realized that when large stars, over a certain size, use up their fuel, they would not necessarily become white dwarfs, which were at that time considered to be the final stage in the life of every star. Combining both Einstein's relativity and quantum mechanics, the boy on the boat calculated that the large stars would continue to collapse due to the ongoing gravitational pressure of their masses. This was the first time anyone had used careful physics and mathematics to substantiate an idea that opened the door to stellar black holes, type two supernovae, and neutron stars. (Black holes had been imagined previously, but lacked rigorous connection to science before Chandrasekhar's equations).

Chandra completed his Cambridge PhD and spent the next nearly sixty years at the University of Chicago. Not an actual astronomer, not a gazer into skies, he was distinguished by his mathematical gift. For decades he was the one man judge running the *Astrophysical Journal*, reading and triaging almost every important scientific paper in the field. Chandra cast such a powerful glow over physics, astrophysics, and the University of Chicago itself, that the University's defining president, Robert Maynard Hutchins, is reported to have said that the best thing he ever did during his long and distinguished tenure was to have appointed Chandra. Chandra's expansive career led to every award available in his profession, such as the Gold Medal of the Royal Astronomical Society, in England, the National Medal of Science in the USA, as well as many other honors. Having initially hinted at the existence of black holes in 1929, fifty-four years later in 1983 he completed his tome, "The Mathematical Theory of Black Holes," and in that same year, King Carl XVI Gustav of Sweden handed Chandra the Nobel Prize in Physics.

Chandra, however, was known for his austere, remote, relentless mind and lifestyle. He worked ceaselessly, and had limited interests and no children. For decades, he was a lecturing professor, overseeing PhD theses, as well as the one-man show at the *Astrophysical Journal*. In the context of such pressure he continued to pour out Nobel level research papers, leading one of his associates to quip, "I have discovered that there is no Chandrasekhar limit."

Chandra was also intimidating and competitive. His biographer, K.C. Wali, quotes the young graduate student, Carl Sagan, describing how students tried to avoid passing down corridors where they might encounter the judgmental professor. Dr. Wali's admiring biography, *Chandra,* nevertheless emphasizes Chandra's driven, and sometimes joyless scientific intensity, which even the Nobel did not slake. Chandra himself confessed that he was not really an astronomer, but a mathematical physicist who probably never looked through a telescope. Undoubtedly, a life without wonder, right?

In 1987, the seventy-seven year old Nobel Laureate and retired mastermind of astrophysics, received numerous invitations to lecture on the three-hundredth anniversary of the publication of Newton's *Principia Mathematica*. The perfectionist Chandra decided to prepare himself for his lectures by re-reading the greatest, hardest, science textbook cover to cover. Newton had once confessed to making *Principia* intentionally difficult in order to slough inferior readers. Chandra described the intense way he studied Newton, and its result.

"I first constructed proofs for myself. Then I compared my proofs with those of Newton. The experience was a sobering one. Each time, I was left in sheer wonder at the elegance, the careful arrangement, the imperial style, the incredible originality, and above all the astonishing lightness of Newton's proofs; and each time I felt like a schoolboy admonished by his master." (There is the child again, and unfortunately in Chandra's case, an admonished one.)

The sense of wonder is often triggered in one person by encountering another person's mind. The overtly stolid but inwardly inspired Subrahmanyan Chandresekhar wrote more about this in his book, *Truth and Beauty*:

"In my entire scientific life, extending over forty-five years, the most shattering experience has been the realization that an exact solution of Einstein's equations of general relativity, discovered by ...Roy Kerr, provides the *absolutely exact representation* of untold numbers of massive black holes that populate the universe. This "shuddering before the beautiful," this incredible fact that a discovery motivated by a search after the beautiful in mathematics (as Chandra perceived Einstein to have done) should find its exact replica in nature, persuades me to say that beauty is that to which the human mind responds at its deepest." He added two Latin mottos:

"The simple is the seal of the true.
Beauty is the splendor of truth."

The wonder that Chandra felt studying Newton, Einstein, and Kerr, an uncanny "strangeness of proportion" that he felt existed between the human mind and nature, mirrored Einstein's own description of the essence of wonder: the awareness and organization of many natural events by a human mind's clarifying insight. Chandra repeated this idea in another way. "It is, indeed, an incredible fact that what the human mind at its deepest and most profound perceives as beautiful, finds its realization in external nature." Chandra borrowed the phrase in quotes, "shuddering at the sight of the beautiful," from *Phaedrus*, by Plato, in which divine memory of otherworldly truth—the enduring Platonic Ideal—is experienced as: "a shudder runs through him, and again the old awe steals over him…"

According to this line of thought, intellectually catalyzed wonder comes from the eerie correspondence between human thought processes and the nature around them that they seek to understand. We shudder at the beautiful in nature because it has some eerie congruence with the internal make-up of our own nature-derived minds.

To express his ultimate intellectual satisfaction, Dr. Chanrasekhar, like Keats, had to fall back on a nature image: he wrote that though he could never expect to understand absolute truth, which is as unlikely as standing on top of Mt. Everest on a perfectly sunny day and seeing the entire Himalayan range in dazzling white snow stretching to infinity, still, "there is nothing mean or lowly in standing in the valley below and awaiting the sun to rise over Kanchenjunga." Wonder does not require perfection, but it is a vision of nature's beauty, whether it is appreciated through eyes or through mathematical symbols.

For the young poet, Keats (who was dead by age 26) and for the aging dean of 20th century Astrophysics, Chandrasekhar, wonder poured into them through cultural mediation. Nature provided the metaphors, such as the discovery of the Pacific or the ascent of Himalayan peaks, but the actual immediate stimulus to wonder was beautiful language or elegant mathematics.

Rachel Carson, for all her love of nature, realized that wonder was conveyed and taught, and is multidimensional: linguistic, mathematical, educative, revelatory, humbling, silencing, expressive. Wonder as we have been tracing it, is powerful, disorienting enough to require re-orientation, and multidimensional, touching more than one aspect of personality. It is emotional, aesthetic, and intellectual.

Chapter Twenty

A New Essence Pervades

There is an aspect of receptivity that provokes wonder. While wonder can occur without calculation or control, and is partly due to surprise, such as the clarity of the stars in the South African night, or the fact that Einstein's equations for General Relativity predicted things as counterintuitive as black holes, still, wonder does not occur randomly either. A certain type of readiness, a certain kind of positioning seems to be involved.

When Walt Whitman heard the learned astronomer, (Chapter Two) he had to glide out of the lecture hall into the mystical moist night air, to look up in perfect silence at the stars. Would he have looked the same, or would he have looked at all, without the initial provocation of the learned astronomer who Whitman appears to disrespect and reject? And why does Whitman describe himself looking up at the stars "from time to time"? Was continuous gazing impossible, or less lovely? Does wonder require this oscillation? Was Whitman's "perfect silence" in the presence of real stars the same as Keats' description of his reaction to Chapman's translation of Homer "...with wild surmise, Silent upon a peak in Darien"? If silence is so important to wonder, then why did both Whitman and Keats break that wondering silence with poems about it?

Wonder is a relational term that describes an interaction. It is not static or one-dimensional. A person "grokking" the stars in a new hemisphere, or reading a masterpiece, enters into a relationship with that "other," and, in the context of that interplay, may experience *the collection of psychological events* that we group together under the term, "wonder." Wonder, like life itself, is a region of interactions. Whitman listens, goes out, wanders, looks, is silent, looks away, goes home and writes. His vitality awakens, and it is thoughtful, experiential, focused, oscillatory, silent and expressive.

Then what makes wonder coherent enough to dwell under the influence of a noun while it retains such life-like variation and motility? If wonder is a blanket term covering so many parts, does it have any central essence at all?

A person in a state of wonder is being bombarded and is holding her reactions together. Wonder is a psychological *holding*, a bear hug of an emotion. Our star gazers in South Africa are seeing many things in many ways, details and panning-of-the-camera, lights and dark space, clarity and scatter, and they are also thinking many types of thoughts, comforting and discomfiting, poetic and harried, scientific and confused. Because the fundamental tenor of their experience is safe and aesthetic, not terrifying or dangerous, they hang in there, like outfielders during batting practice, shagging fly balls one after another, except that, in their case, they are not only catching balls hit at diverse angles, but also arriving from different dimensions. They are more like Einsteinian baseball players catching balls hit in four dimensions. Their sense of wonder is hugging together sensation, emotion, thought, and experience. Psychologically, the correct word for this pattern of mental life is "integrative." Wonder is not an entity or a fixed state of mind. It is an integration of many psychological functions.

Wonder is an integration of body, mind, and feeling, through qualities of receptivity and attention, a synthesizing and wholistic apperception. Of course wonder is inside the person's psyche, but it is relational, about something, provoked or triggered by something. So here is the crux.

Wonder is a set of disorganizations, integrative organizations, and reorganizations, in response to complex stimuli that are multidimensional, such as the night sky, great poetry, or scientific breakthroughs.

But not every reorganization is wonder-full. In fact, many modes of re-organization are aimed at reducing wonder. Wonder can be stifled when an experience is re-inserted into, and held in place by a routine narrative. ("The stars were pretty last night." "Einstein's formulas uncannily predicted black holes, but, hey, Einstein was a pretty smart guy.") Wonder describes two additional features. Its integrative holding *does not reduce the new to the known*. Something new is absorbed and retained. There has been a psychic change.

In addition, we tend to reserve the word, wonder, for the receipt of wholes. Wonder goes beyond the mere addition of new facts, ideas, or feelings. In a state of wonder, there are innuendos or echoes of presence. An old way of knowing shatters, and *a new reality*—not merely new facts or events—but a new essence pervades.

For example, Keats' poem implies more than just the geographical discovery of the Pacific. On the far side of the Americas lies the biggest ocean in the world, the Pacific. The innuendos of Keats' poem are that not merely a new place, but a new concept of planet Earth has been found: the world is very much bigger and different than Europeans had previously supposed, and this becomes a metaphor for the fact that Homer not only wrote epic poetry, he inhabited and evoked in literature a new world of human experience.

Or, Chandra seems to be saying, Einstein not only did some fancy math, but he also envisioned the universe in its totality, and described it in symbols whose intricate, detailed, unraveling leads to precise confirmation in uncanny detail. The products of human thought, like Einstein's formulas, strangely correlate with cosmic realities, and this correlation, as much as the realities being described, evokes wonder.

Or, according to our North Americans in Africa, the stars in the zone of the Southern Cross can carry us beyond anything we believe in, a sublime and inexplicable beauty, our night of nights.

In each of these brief summaries of more expansive moments of wonder, newly revealed details catalyze new whole psychic realities, not merely additional information.

Chapter Twenty-One

Narrative Urgency

I n wonder's thrall, something is recognized as present. Something, an "other," emerges from the mere machinations of materials and natural laws. Wonder's reception of a new essence is built on the human relational capacity, so that wonder feels connected-to.

In rapt attention and delight, a presence is felt to such an extent as to evoke ongoing involvement, and the desire to understand and relate more, to preserve or follow on the trail of wonder.

Wonder retains from our mammalian biology a quality of "imprinting," such as when the ducklings whose first sight, after hatching, is the scientist, imprint to, and forever follow that Austrian PhD as if he were their mother. Biological evolution has designed newly hatched ducklings to trust and depend upon their initial "other," which, with the exception of human-designed laboratory experiments, is destined in fact to be their ducky mom.

We say we felt wonder when we also feel imprinted, and there is an *enduring, memorable salience* that establishes direction. If we felt wonder, we follow it. We have been imprinted to direction-giving relationship. Like a duckling to a duck, like a child to a parent, like a scientist to a new galaxy, wonder contains a sense of invitation.

Wonder is a recognition of emergent complex wholes that are felt to be living. Wonder is embedded in relational modes and in subjectivity as well as in knowledge, and includes feelings of presence, kinship and connection. Wonder is for children, and for adults who once were children and who felt a gratifying and uplifting connection to the world around them. Wonder is psychological, intellectual, cultural, but also social. It is a form of dynamic connection such as we feel towards our family and community, and this implies something about communion.

That brings us around again to the oscillation of silence and speech, why Walt Whitman listened to the learned astronomer, but then left, and looked up at the stars from time to time, and kept perfect silence, which he later broke to write a poem. We have seen that wonder is taught and communicated, and that means someone is also expressing it, Rachel Carson to her nephew, Whitman to his readers, Newton, somewhat begrudgingly, to generations of math-competent scientists. Generally, in wonder's thrall, we fall into a silent, wonder-struck receptivity, a standing back, a flooding in, an internal process of disruptions and restorations, a wide-picture realization, like an aster in sunlight.

But Keats gaped in wild surmise only so long before he was penning another sonnet. We generally experience wonder as infectious, something we want to show-and-tell about. In biological language, we could say that these experiences evoke *recruitment*. Like ducklings we want to follow wonder, and like crows we want to call it out from the treetops. Wonder is something you want to convey.

There is an urgency to tell others what is important to us, which has its roots in primate adaptation. When the five-year old returns from kindergarten, and pours out non-stop the details of her day, she is storing information, integrating it, rehearsing solutions to unsolved problems, re-living delights, and also eliciting commentary and education from her attentive mother. An adult who has felt wonder will re-experience the kindergarten child's narrative urgency.

Wonder is something your group of cavemen may well need to know about, but more importantly, wonder is something you must disgorge in order to not be choked by it, for in its essence it is a vision-quest moment of a fulfilled and therefore charismatic social mammal. The rest of us need to know what you have witnessed. What you saw and felt in the night sky, or while reading your treasured sacred text, belongs in our global positioning system, belongs in our myths, belongs in our aspirations. The crow has found meat and screams it to her hungry mates; you have found wonder, and you want to put its colors into words, and lead your tribe back to the promontory from which those golden rays could best be seen. Wonder is also *a social emotion* that extends like a golden thread from its inception through its telling.

The imprinting and recruitment that wonder evokes reminds us of that elusive gray zone between awe and wonder (Chapter Eighteen).

Now we can add that awe is something to which we do not feel imprinted. It is wonder which, along with its flooding in, its disorganizations and reorganizations, and its salience, also establishes a confirming and secure emotional tone. Is it possible that wonder is often described as a property of childhood because one of its attributes is association with the feelings we once had towards our own mother? Would Freud have something to add about wonder and mothers?

Freud considered religion to be a regression back into feelings that had been previously felt in childhood. He wrote, in *The Future of an Illusion*, that we are afraid of nature, of our bodies, and of death, and in reaction, we imagine a fantasy in which we are protected by a good father, as once in the past we were children under the safety and guidance of our real fathers. (See Chapter Twenty-Eight). We have already seen how wonder is recurrently associated by Rachel Carson, Dylan Thomas, or Albert Einstein, with childhood, (Chapter One) but also how it may be associated with cosmic destruction and all devouring time, as in the *Bhagavad Gita* (Chapter Two) or even Carl Sagan's *Cosmos*. Is wonder a regressive, childlike feeling, an attempt to slake unanswerable fears with parental soothing? Does wonder imprint us to a false mother as happens to ducklings in ethology experiments? Might wonder be another flavor of reality-denial?

According to Freud, the hallmark of *pathology* in religion is obsessional ritual, repeated self-abasement, and divine propitiation, plus superstitious correlation between ritualized behavior and supposed outcome of future events. Wonder has none of these features. Wonder does not fixate on one authority, one interpretation, nor upon intentional preservation of infantile helplessness. Wonder shows no relation to obsequiousness. The feeling of childhood connected to wonder is a temporary emotion and not a ritualized fixation. The image of a benign parent does not occur in wonder, though emotional association, without the image, may be present. As we are seeing, emotional association to many aspects of mind are integrated in wonder. Nor does Freud's description of religion necessarily hold true for all of religion itself. Freud's insight is provocative and students of religious feeling would be wise to take it into consideration, but it is not necessarily conclusive or most often true.

Still, it is interesting to recognize in states of wonder the arising and passing of childhood's feelings, and parental evocations. These tell us little about regression, and remind us instead of the integrative holding of many aspects of mental and emotional life that accompanies wonder. Wonder involves uniquely human dimensions of imprinting, family life, development, and communication.

Human imprinting, unlike the avian modality, involves a questioning, an aesthetic, a quest, leading us on as the initial moments of wonder fade away. We don't suppose that Keats put down the volume of Chapman's Homer, and then never got around to reading that book again. More than fifty years after the brilliant Indian boy on a ship bound for England first intuited how stellar masses and gravity could produce phenomena like black holes, Dr. Subramanian Chandrasekhar was still studying and mathematizing this wonder-working cosmic incongruity.

So far we have developed a portrait of a human capacity that we call wonder. It is a response to multidimensional stimuli that are salient, disorganizing, and that demand our reorganization, that are accompanied by a new source of reality that integrates multiple psychological functions, that give us a new sense of direction and a secure emotional connection, and that we want to shout about to the heavens and our friends.

The imprinting, social, aspect of wonder raises a problem that we have to consider, because wonder in some ways resembles, but in other ways differs from interpersonal relationships. Of the thousands of people we encounter, how many evoke a sense of wonder? To consider the limits of the communitarian properties of wonder, let's imagine that we are taking a walk on the seashore. There we will see something about wonder that challenges its sociability.

Chapter Twenty-Two

Leaping Out of Categories

A man is walking along an Arabian Sea beach in Southern India. The shore is lined by coconut palms whose fronds are pierced occasionally by paler green minarets rising from whitewashed mosques; and at ground level, cattle bathe in pools, and teams of fishermen tug-o-war long heavy nets. The sun is setting over the ocean, which rises and falls leisurely, pocked by ripples, each one of which catches golden sunlight at a shifting angle that sparkles. Far out towards the horizon are tiny, one-man, coconut–trunk dugout catamarans, bobbing toward the liquid edge of the world. Antiquity and languorousness mingle with the emptiness of treacherous water. The man thinks:

How long has this benign evening been repeating itself? When will the sea reveal its tsunami shadow? Will the tiny fishing boats be warned of any impending danger ahead of time? How many ripples dimple each wave, this expanse, the ocean? How many ripples have arisen and passed away on all the oceans since this ocean was formed when, millions of years ago, continental drift split India off from the ancient super-continent, Gondwana, and smashed it into Asia? Is there a knowable number large enough to enumerate the ripples I now see, all the ocean's ripples right now, the history of all ripples? Are ripples a telling analogy for the brevity of my life? It seems as if some mood pervades this Malabar Coast; is it my imagination? If I stayed here, what cosmic atmosphere would enter into me and what would I come to know? How did I end up here? Were there really no humans on this coast only fifty thousand years ago, before the homo sapiens' migrations out of Africa reached as far as India, when the sun, palms, and water created the same mood, but on a lonelier and more naked planet? And two million years ago, before homo sapiens, this sunset spoke only to a planet without poets and words? How did the world ever become so beautiful, or is my sense of beauty only a conditioned, subjective perception? My seeing of this golden dusk feels as creational as the events themselves, as if my seeing were itself a part of the causation of this scene. My presence here feels propitious and destined. Yet everything is as transient as the ripples on the swells. The sun is going down; here are its last rays; the Earth feels warm like a bosom. I feel eternal and ephemeral. I feel I am Earth's appointed witness but also one fading light. How can I feel simultaneously so full and real and insignificant?

Wonder may flow out from a sense of an other, but it also involves "non-reducibility." One feels something stretching thought's joints, something encompassing, incomprehensible, and beckoning. The non-reducibility is not a product of an information deficit, but an experience of the insufficiency of all explanatory systems. Wonder pours over the banks of categories, beyond answers but not beyond questions, precluding on the one hand convictions, and on the other, cynicism. Analogously, Lewis Thomas once defined science as, "elicitation of disbelief and celebration of surprise." Wonder is not a duck, not a mother, not a pretty scene. Wonder is disorganizing, reorganizing, a social message, a wholistic relationship, an imprinting, and an experience that leaps out of categories. It is as if wonder were round and could not fit into any box.

Wonder shatters conviction, with its attendant self-righteousness and self-appointment, and accordingly contributes to beneficent human concourse. Imagine how differently history would have evolved if every Inquisitor or Jihadist could tolerate wonder, instead of forcing personal conclusions onto others. Who ever heard of a sense of wonder driving hatred, or conquest? There may be an analogy for the relationship of wonder and conviction in the Heisenberg uncertainty principle. As was discussed in Chapter Seven, the uncertainty principle has demonstrated that some forms of information immediately banish other forms of information from our world. Knowledge about an electron's position immediately eliminates knowledge about an electron's speed and vice versa. Similarly, wonder and conviction are not merely opposites, but seem to dissolve or cancel each other out. When you believe strongly, then you feel you know, and wonder evaporates. Wonder is stored in a psyche where the files aren't filled and closed. It endures as a benign stimulus to questioning and re-assessment, elusive as well as attractive. As soon as it is caught and pinned down, its life force escapes.

We call it wonder because of its numinousness. Reduce that to conventional explanations, and you are left with a pleasant evening on the beach.

Wonder is a psychic pregnancy, life-giving and validating. But it is also asocial and disconcerting.

We feel wonder when we are reaching out to maximum dilation at our extended information horizon. When we say we felt wonder, we mean that we have been expanded, in an important and enduring way.

This degree of open-heartedness arises in places and moments where *meaning and meaninglessness are both present* and pulling on the big rope that connects them. An example of this is when you are looking at the ocean with both delight, and also with a sense of personal vulnerability. The feeling of connection to an integrating presence pulls one way, and, tugging you in the opposite direction is the loss of explanation and answers. Take away either of the rivals, and wonder loses its unique location of comforting feelings amid disconcerting vistas. This tug of war expands the mind's reach in opposite directions, and helps us understand why Whitman looked up at the stars "from time to time," or why the stargazers in South Africa felt both uplifted but exhausted. Wonder is not always cozy, maternal, and warm as discussed in Chapter Eighteen; wonder edges towards awe. Wonder may include but is by no means limited to the maternal. Wonder simultaneously contains "meaningless," "alone in an empty universe," "wheeling with the stars," "silent in wild surmise," "a gift to each child in the world," "Beauty is the splendor of truth," "the sublime cradle."

Let's look at another example of this horizon-dilating dialogue implicit in wonder, about integration versus anomie, such as might occur at a wild and windy edge. We are discussing an ability to hold in one frame opposites that are racing away from each other.

Chapter Twenty-Three

Primary Encounters Beyond Concepts

A woman is walking on a cold, windy Cape Cod beach. The ocean is steely, frothy and violent. A brief dip in it might pull your body temperature below survival level, or drag you out into the impersonal churn. Big brown seal heads rise up like dogs from an alien medium, watching the watcher. The seals' eyes are like periscopes of the deep. In the distance, gannets dive hundreds of feet with enough gravitational momentum to penetrate the icy depths and catch fish. Small flotillas of duck-like eiders and scoters float beyond the breakers, ironically comfortable and relaxed in the numbing water. Going outward, the next stop might be the Grand Banks, Arctic seas, or British Isles, distant destinations across cold waters that can so easily snap the warmth of human life; and yet these highways have been traveled by fish, whales, birds, and people in hand-made boats, from time out of mind. The stormy Atlantic itself is a recent invention of the Earth, a seam in the garment of continents that opened as continental drift tore the super-continent, Laurasia, apart in the days, tens of millions of years ago, before Earth had thinkers or self-awareness in any form. On frozen edges of this ancient anarchy, only ten thousand years ago, humans chipped spear points. As canny as these migrant hunters must have been, how their thoughts must have differed from ours! In that risky, seasonal, life-and-death world, how they must have stood, spear in hand, poised above hunger and sudden death; and how, at last, they all disappeared anyway.

Now the full moon rises reflecting light from a sunken sun. Gannets plummet into the silver water, gulls and terns wheel, and petrels flutter in the lunar distance. It took tens of millions of years for these bird species to form by evolution out of their ancestors. It took billions of years for Earth and moon to form from their mineral constituents in a solar system strewn with the refuse of their exploded ancestral sun. Waves lace the Cape Cod shore, dancing to the Earth and moon's gravitational beats.

The woman walker tries to hold in her thoughts bird evolution, moon rock congealing in the solar system, the mind watching from behind seal eyes and her own eyes, the mixture of law, revealed in gravity and adaptive evolution, with randomness, revealed in waves of the unquiet ocean, itself full of concoction, chemistry, proto-life. But what she sees and feels cannot be held together by her scientific thoughts. The welter of her experience is a criss-crossed traffic of Earth history, biological evolution, human destiny, aqueous caprice, her own receptive sensorium, the smell of salt water and the pewter sky light.

Here is a world in which we feel simultaneously embedded and abandoned, a world of familiar beauty, and of forces and histories that deny us any role or purpose. As meaning and meaningless contend in our mind, in their tidal back-and-forth, wonder grows from recognition that we are born out of and cradled within an immeasurable cavern of immeasurable complexity. Wonder is impeded by boundaries, limits, beliefs, and doubts, and is facilitated by *primary encounters beyond concepts of time, space, self.* As long as you can hold on clearly to what you are thinking and feeling, you are not in the state of wonder. As long as you have clarity about the meaning of what you are experiencing, you are not in a state of wonder. Wonder does not rise up merely from confusion and doubt, but it does occur only when there is a flooding and blending of many aspects of mind, including thoughts and emotions. Wonder is a zone rather than a point.

John James Audubon, a man of wonder, is often portrayed as an indefatigable nature lover, which of course he was to a very large degree, but he experienced wonder's tug of war, and could also be repelled by nature. He describes an experience: "...as if I were almost upon the verge of creation where realities were tapering off into nothing." In order to spend years of wonder in nature, one also has to peer beyond its pretty and pleasing edge. Wonder is fertilized by mixing together both knowledge and its limits. Wonder is the beach between home and drowning.

Wonder is not an entirely relaxed holding state, not a mindless blankness nor a relaxation response. It contains the tension of complex awareness, just as alert meditation uses postural muscles to create optimal wakefulness. There is some mental work involved, a suspension bridge of mind. It is like the light tension that maintains the soap bubble, spinning and refracting its myriad tiny rainbows as it rises. A person in wonder may be rotating with rainbows, but is also somewhat stunned, like artist

Audubon teetering at the edge of nothing, like astrophysicist Chandra, "shuddering."

Where a person is not under threat, and allows themselves to receive conceptually, emotionally, or aesthetically, the world in all its complexity, they are in a good position to open out into wonder like a plant benefiting from all of the signals and juices in rain, sunlight, and soil. In answer to a question which was raised in Chapter Five, we will now clearly see the evolutionary, adaptive component to wonder, and we can assign a reason to its preservation in the human psyche.

Chapter Twenty-Four

Irreducible Complexity

Wonder is the way that our ancestors, expanding out of more limited, early primate minds, allowed themselves to receive more input, to absorb information simultaneously from multiple channels. Wonder is, among other things, adaptive information-processing. It produces both salience and a subsequent exploratory drive. It functions as a catalyst for observing resources and facilitating fresh responses. It focuses attention on multiplicity and processes rather than on items. Tactical and strategic interactions with the world contain lower information-value and may preclude further attempts at understanding. We expand and glow with wonder as an adaptation to live with *greater receptivity, greater awareness, less preconception,* and less turning down of the dial. Wonder is wide-angle information reception, a sensitized rotating "dish" of sensorial and psychic signal capture. We have evolved to contain within our menu of options a *maximally absorptive state of being.*

Wonder is our own complexity kindled by the complexity of the world. We need to awaken and utilize all of ourselves to be so widely receptive to so much.

We are talking about moments when all of us is engaged with all that is around us. The coordinated complexities of our human nature are called up into use. The complexities of our environment are recognized simultaneously. We feel in a relationship; we feel connected to a warm presence; we feel emotionally wide open; we feel our senses absorbing; we feel our intellect awakened and challenged; we feel alive and called into the world with a meaningful purpose, and we feel evaporated and unable to assign any meaning to the enormity of our experience in reference to so much "other."

The more we know, the more we recognize what we don't know, the more we are eager to know, the more the rain of information soaks our leaves and petals. "Information" means not just facts, but colors, energies, patterns and people. Wonder is the experience of irreducible, overwhelming complexity, that feels important, relevant, based on sensory input that is flooding over the banks of categories and washing away old beliefs and assumptions, yet which can be sustained, held up without diminution, not destructive, and which feels predominantly positively transferential and beneficent.

As eyes evolved to process the information in light, *wonder is the receptor for complexity*. The word "receptor" here is intended to imply that wonder is built into us. We are born ready to receive it. As we have a preexisting apparatus to receive sound waves in our ear, or light in our eyes, we arrive on Earth with a pre-existing matrix to capture the resonation of wonder.

In Section III we will examine in detail how the life-system for wonder-reception arose: pouring into us from the laws coded into the universe, or from "the initial information state within the Big Bang," from the evolution of our galaxy and of our sun, from the photons of sunlight which contain the energy and directive information that has been stored in the hydrogen atoms of the sun (and which were released to jet across the sky to us at the speed of light from hydrogen-fusion reactions in the sun's furnace) and captured by the receivers in green plants, comes the energy and the guidance by which atoms are arranged into organic molecules within the cells of life.

We can never stand outside and neutrally observe this electronic, thermal, chemical information system of the universe, which formats life on Earth, because we are of it, a product and derivative of it. When we *open out in full relationship to this cosmic complexity we feel wonder*. When we consider its duration of billions of years, its trans-galactic expanse, its compounded order accumulated in the subatomic world, its opaque origin; when we allow ourselves to absorb its aggregated information storage in its "forms most beautiful and most wonderful," about which we are unable to formulate any credible statement of origin, goal, or end, then wonder becomes our wholistic psychological holding mechanism.

We are talking about the slowly roiling mind, as it integrates the flow of energy that started with the cosmos, and that is passing right now through our own bodies and environment. The boundary-crossing span of this ceaseless, energetic transformation of all things is the stimulus to wonder. We experience ourselves within the cosmic flow of matter and energy. We are both a product of it, and able simultaneously to observe it and realize it. Everything is moving and changing in interconnected patterns.

Wonder is an adaptation that exists because complexity exists, and because our awareness of that complexity facilitates our flexible and creative responses to the world.

In the lawful universe, order, chance, and history accumulate into the patterns of phenomena around us. Our bodies, and the surrounding world, contain aggregated, interacting selections, asymmetries, paths, and arrangements of quarks, electrons, atoms and molecules, heaps upon heaps of directed and constraining arrangements, at astronomical, cosmic numbers, and vibrantly influencing each other. There is a physical arrangement of the world that is a product of impossibilities and necessities, and we call these forces "laws of nature." They express information that is impregnated in existence, and that guides it fast and loose, forcefully and lackadaisically, directing supernovas, dinosaurs, and humans. We are embedded in complex, compounded, ancient history. We experience wonder with our mind that is itself the outcome of the very process that provokes us into wonder. Complexity at the magnitude of quadrillions and quintillions cannot be understood, and our minds make a phase shift from information to wonder.

Wonder is *complexity self-aware, complex awareness kindled by complexity received*.

Wonder is not just about complexity; it is itself complexity.

Wonder is our way of remaining connected to experiences that exceed other relational modes. When an experience is beyond ordinary thought and feeling, beyond ordinary comprehension, we might be able to hold it in wonder. It uses our skill in staying in relationship, and applies that skill to its maximum expansion. Wonder is the door that permits us to experience irreducible complexity.

We can face and hold within us our source.

Chapter Twenty-Five

Mirror of Cosmic Magnitude

Wonder is everywhere. It is around us and in us.

It frequently happens that when a child is born, wonder scintillates through the room and penetrates into every beaming witness. Like Brontislaw Malinowski's anthropological subjects in remote oceanic islands, who, it used to be said, did not causally understand the connection between sexual intercourse and subsequent pregnancy, at the moment a new human appears, we are often unable to connect yesterday's hug and flash with today's fully formed future president. To many a new parent, the radiant little entity magnetizes their emotion and defies comprehension. In the eyes of the new being are mirrors that understand. How did emotion and understanding enter into a lump of clay? How did all that pizza and ice cream, or even the healthy lima beans and chicken, that the mother ate, become circulating blood and perfect little replicas of toenails? Many a young doctor and new parent has felt his or her lectures on cell division and embryology evaporate in risible inadequacy in front of a *totality* who defies the mere additive summation of miniature parts.

New little people seem to have skidded down some great invisible water-slide from a somewhere else. When we attribute this phenomenon, life, to the information coded in parental genetic DNA, we end up diminishing a phenomenon rather than explaining a fact. We think we understand what we cannot actually comprehend. DNA is crucial of course, but it is far from being a full explanation. We laugh and cry in wonder, because we intuit that, even given the billions of information bits in DNA, and given the four-and-a-half billion years of evolution of DNA on our planet, these alone do not account for our new child. DNA function itself requires that perfect organizer of particles, the electromagnetic plus-and-minus force. The Earth, on which DNA evolved, is itself a child of gravity, so that, like Carl Sagan's apple pie, it takes gravity, stellar evolution, planetary creation, an entire universe to make that baby's wrinkled, holding fingers and her mother-melting smile.

And, as far as we know, it takes a child-grown-up to see the universe. We have not yet encountered any conscious awareness to rival the human. Humanity emerges into the world, Tagore said, "like a bud in the forest at midnight," containing adequately complex receptivity to be a scientist, artist, mother, or wonder-struck child. As far as we know, it takes the universe to make a child and a child to understand the universe. All of those children of wonder who we met in Section I—Rachel Carson, Dylan Thomas, Albert Einstein, Carl Sagan, Rabindranath Tagore, Pablo Neruda—are remarkable not only for what they thought or wrote, as scientists or poets, but for the constellation of receptive awareness they contained, and which to some degree, we all contain.

The universe has designed a mirror of cosmic magnitude that can fit inside a damp little head with a soft fontanel.

The universe seems to have produced something that can know it. That "something," us, contains not merely the biological information-ordering of DNA, but all the laws and forces of the universe, from electromagnetism and gravity on down. In all of our trillions of cells, all of the forces of the universe are operative.

Intricate, interlocking and marvelous, the information that has been aggregated into us during Earth's development is called up inside us, and mirrors the scintillating world. We feel most fully alive when we experience ourselves as manifestations of the laws of matter, the laws of sunbeams, the laws of life. They all intersect and weave us. We are situated within and aware of what Harold Morowitz calls the "combinatorial explosion." (Chapter Twelve) Huge numbers of small parts are combined within us. The units form combinations that become units and combine, in a cascade of interactions that reaches from atoms to minds. Wonder accompanies any inkling or intuition of this. Inside of us and as our tailwind is the system of the cosmos. When experiences evoke this aspect of who we are, we delight in wonder like hummingbirds iridescing in sunlight.

Chapter Twenty-Six

Pre-Copernican Minds

But here we have to pause, and recognize how wonder can sometimes be drained by our tendency to self-flattery. Our need to feel important can distort wonder into spiritual sentimentality. Though our baby was made by the cosmos, the cosmos was not made for our baby. Our pre-Copernican minds, which insistently imagine that we are located at the center of the universe, are never far away, with their demand to restore us to center stage. Magazine racks at the health food store sport headlines of neo-pre-Copernicism. ("What does the universe intend for you?"). Once again we have tried to make ourselves the center of the world. The Middle Ages have not yet fully ended.

The more we learn, the more it is appropriate for us to be filled with Whitmanesque expansive wonder. We are products of an ancient, possibly eternal creativity, from which we derive, and which we contain. But wonder is not the same thing as personal self-adoration, or cosmic hubris.

As we have seen, wonder holds and integrates opposites and complexities, so that, with inevitable irony, it can survive only in an atmosphere that acknowledges its own limits. Even science, when it is held up as a source of all answers, can destroy the sense of wonder. Wonder is compatible with a causal universe, but not with dictatorial reductionism. Neither as scientists nor as theologians can we claim privileged access to statements about the universe's origins, intentions or goals. We can't vault ourselves to an observing station above the space-time from which we derive. Our perspective is internal and parochial. Knowledge of the meaning and end of the universe would require measurement from an external platform.

No reasonable person will claim to understand a goal or purpose for the quasar-studded, and black-hole booby-trapped universe.

Our exquisite mirror of the universe, the mind of our baby, and Hubble, Einstein, Whitman and Neruda, which we all culturally inherit, and which we ourselves may also contain, does not provide perfect resolution. All mirrors have imperfections. Just as we can see some of the electromagnetic spectrum, which we call "visible light," but can't see most of it, like the infrared and ultraviolet rays, similarly, we may well be entirely blind to aspects of reality. We who float in the cosmic ocean can't describe its far shores. Even our little garden bumblebees can see parts of the light spectrum, and therefore nuanced floral colors, to which we are hopelessly blind. Our telescopes cannot exit from within the Big Bang's cosmic egg. If there is a multiverse, or a series of expanding and contracting Big Bangs and Big Crunches, we are likely never to be able to differentiate hypothesis from fact.

Experiences of wonder leave us feeling as Einstein did, both reduced and augmented, humbled and freed-up. Wonder helps us admire the theatre set, but we contain common, coarse emotions that seem to demand a denouement before the playwright has spelled out the entire plot. How can bit actors in an epic trilogy claim to be the playwright's chief expositor? We are cameo players in a drama whose heroes include plate tectonics and supernovae. Our ability to trace the gossamer of wonder is shattered by the bad breath of our self-important bombast.

There is something in human nature that wants us to play first string and refuses to be sidelined. Maybe this is a life-preserving trait when we are children whose demands for attention make us recipients of the parental nurturance that we require in order to thrive. But when we insist that the universe adore us and pay attention to us and only us, wonder gets shelved, and out come beliefs, ideologies, and fact-free, pseudo-explanations. Wonder scintillates when we can remember that we are children of the universe's laws and history, but that all our evidence cannot prove that we are inevitable, destined, nor intended.

Chapter Twenty-Seven

Seeing Clearly in All Directions

To live a life of wonder, we need to modulate our over-confident cosmic narcissism. It seems that no matter how clearly we comprehend science and poetry, we also harbor a tendency to regress from knowledge to wish. By "cosmic narcissism" I am referring to those moments when attention is allegedly focused on the sense of wonder, but the sub-text refocuses attention on oneself. We need to observe the difference between the thunder-struck appreciation of life, and self-infatuation. A good case study for this is dinosaurs.

Dinosaurs dominated Earth for over one hundred million years, a lengthy experiment, before they were assigned to the same trash bin as rotary phones. Paleontology has revealed an Earth history of many periodic mass-extinctions, such as the one sixty-five million years ago that erased almost all of the dinosaurs. Based upon the studies of Alvarez and Alvarez, a father and son scientific team, (the father holding a physics Nobel Prize), the extinction of the dinosaurs is now generally believed to be due to a collision between Earth and an asteroid. (Some alternative hypotheses also remain credible). The massive collision triggered dust, darkness, changes in climate and ocean levels, lasting many lifetimes, and large creatures with exposed skins, thermoregulatory problems, and large caloric needs, vanished, while smaller, subterranean beings, like the new small, furry mammals, squeaked through. Whatever the cause, we know that our planet can elaborate giant beasts, giving them full play and central stage for eons, and then erase them in a flash. Neither determining laws, nor the slow evolution of life by gradual adaptation to conditions, can explain the history of life, which includes happenstance and catastrophe, and does not walk a straight-and-narrow line. What possible direction, or intention, could account for such a long and complicated blind alley as the dinosaurs? Yet there have been even bigger extinction events multiple times, earlier in the fossil record. As we saw in Chapter Eight "...extraordinary events can happen without extraordinary causes."

This is how Stephen Jay Gould (Chapter Eight) and his colleague, Niles Eldredge, challenged the Darwinian idea of gradual evolution, and replaced it with the concept of "punctuated equilibrium." Life on Earth sometimes evolves through step-by-step adaptation, and sometimes explodes with catastrophe and revolution. Nature includes the path of a fly, the cricket's chirp, volcanoes and renegade asteroids. We have spilled out of a cornucopia of cause, coincidence and collisions. We live in a world of laws and loopholes. We adapt to conditions, but the conditions themselves change, sometimes drastically. The important point is that directional intention is inconsistent with the world around us, the world that spent so much time and effort on dinosaurs. (In Chapter Eleven we discussed how chance and history can become determined paths that get incorporated into law). The dinosaurs were neither intended nor unintended. They served no goal or final purpose. They evolved as remarkable adaptations to conditions, and when the conditions were shattered, they too disappeared. Doesn't that reflect on us?

But then again, our world is not a zone of caprice or anarchy. Even chaotic accidents, like asteroids smashing into Earth, occur within constraints, like the laws of gravity, or of entropy. (Chapters Thirty-Five and Thirty-Six) With stellar explosions and planetary extinctions, we see that there is no one tendency to order. But within our own bodies, and the leaves of trees, we do see massive order compounded. Order is not an ordained direction, but it is an emerged and inescapable presence. We do not know whether the universe *intends* to make complex, conscious life, but we know that it *has done so* and that we are here. We are one possible outcome in a universe of law and chance. To think clearly, we need to consider both the massive ordering of molecules necessary to make a single cell, as well as catastrophic extinction that occurred to the dinosaurs and to many other creatures before and since. (The issue of cosmic "intention" will be discussed in detail in Chapter Forty-One).

The universe is a creative tinkerer. There are roads taken, roads not taken, or roads turned-around-on. Evolution reveals many dead ends, creatures preserved in old earth sediments who did not contribute to subsequent life. In the past 500 million years, most of the world's life forms have simply ceased to exist. Scientists estimate that ninety-nine percent of all species (not merely individuals) that have ever inhabited Earth are now extinct. These other extinction events were probably due to causes other than asteroid impacts, such as volcanically induced global warming, or the floods and catastrophes that accompany the break up of continents and the creation of oceans. Many less dramatic extinctions occur when a species is out-competed by others and forced off the stage.

Almost all of life is not a part of the ongoing development of life. Most lives, even most entire species, pass their days in blind alleys and dead ends. Most of life does not contribute to the evolution of life; it lives, dies, and leaves no legacy. The ancestors of contemporary life are the chosen few, adapted to their historical moment, and capable of mailing their DNA to the next generation.

The sense of wonder is often stifled by belief systems that preach destiny instead of the cosmic mix of uncertainty and creativity that is the source of our wonder. The universe is creative for the very reason that it is not predictable. We are struck with wonder at the fact that we rose up dripping from the womb of the universe, that the universe's coding guides from within us the energy and matter of which we consist, and that we are also products of long lines of historical conditions, probabilities, luck, extinctions and holocausts. When it comes to an individual life, our planet is reckless, profligate, and probabilistic. Many are born, few are chosen, and all are traded in for new models.

If we are the children of Somebody, it is of a gambler, a free-spirited, inebriated artist, a wastrel rich-kid with excess to spend. Out of gravity and photons there is us. There is order, but not a plan. There is information, but neither a direction nor a conclusion.

One has only to think about the human propensity for destruction, from the Spanish conquests of Mexico and Peru, to Auschwitz, to the Congo in the 20th century, or to the buffalo of the American West, or to the Blue Whale, to see that our lives are not following an ironclad, upward-bound pattern. The evolution of our planet resembles the oscillating probabilities of quantum physics more than the fixities of progress. We are transitional, perhaps experimental. We are what has happened. So far. Holding our existence in wonder is not the same thing as proclaiming that we are, or understand, some purpose, goal, or ultimate meaning (Chapter Forty-One).

Within the web of cause and sequence, that is strung over the black eons of an expanding universe, and that includes continents adrift and trillions of lives come and gone, many causes, forces, and accidents are woven into the pattern. There are many dead insects on the screens of time, lost in a cosmos of chaos and violence, and there are the origins of intricate, delicate, ancient order that can be felt by us from within us.

With the human organ of wonder, we feel a web of regularity impregnating ourselves and the universe. It is very old, very complicated, and very profligate. Its scale and its style are not modeled on our way of doing things. It continues to build, break, renew and tinker. Sometimes whole galaxies are deleted. Sometimes stars, planets and billions of lives are erased. And through it all, a great deal has emerged, and here we stand, inheritors of Neruda's verse and Hubble's telescope.

Wonder is the state of mind that evolved to hold this all aloft, to bear in mind what cannot be concluded. It is a psychological database that includes facts and feelings as valid information. Wonder, like the series of prime numbers, is an expanding set without limit. When fear, poverty, authority, or conformity do not constrict us, and when we look deeply into the lights in stars and baby's eyes, wonder is kindled in our brains as a response. The lights in the universe turn on our own lights with luminous dilation. Wonder is the courage to see clearly.

Chapter Twenty-Eight

A Full Rainbow of Effect

I deas are an adaptation with which our minds can contain useful information, but, in contrast, wonder is an adaptation by which we can sustain awareness of complex, non-reducible wholes. A whole baby, a whole universe, a whole sunset over the horizon-rimmed sea, may provoke the feeling of wonder, and thereby stay alive in memory and association, in their full echoing rainbow of effect. Wonder includes the ability to not-know, to not conclude, and to not-deny, like the psyche's skillful juggler, keeping many balls aloft. Not only balls of ideas, but balls of impressions, moods, and intuitions—so wonder may be like those jugglers who can keep aloft balls, bowling pins, and flaming sticks.

But this raises an inevitable question. If wonder is so beneficial, pleasant, realistic and life giving, then why does the human community so often appear so wonder-impoverished? This book might be read as a paean to wonder, and an exhortation to promote it. But if wonder is so obviously a blessing, then why does it need an advertising agency? Why do billions of people feel convinced that they are privy to some unimpeachable, ultimate knowledge, that gives them the right to burn alive, behead, imprison, or silence their unconvinced fellows? Why do so many people seek the comfort of conviction instead of the delight of wonder?

On the big scale, we see belief and bigotry as major currents in the river of human history. Wonder has never been seated on a horse at the head of crusades or jihads, or been a platform for election. ("Vote for me! All of my proposed policies are filled with questions and wonder!") On the local scale, why are so many of our conversations with ordinary people so emphatically tinted and so conclusively punctuated? Ms. K. knows the cause and cure for cancer, and Mr. J. knows how to bring world peace and fix the nations' schools. In what conversation today have you encountered a person simmering with reason, love, and intuition, musing on the relationships among things, persistently curious and inconclusive, inspired by the polyseminous concatenation of the world? Haven't you found yourself encountering predominantly unknown, unappreciated, self-appointed presidents and popes?

Clearly the sense of wonder is suspended in dynamic tension with other human needs, such as certainty and security. Particularly for people whose lives are overwhelmed with fear and insecurity, the sense of wonder may be too psychically costly to maintain. The woman with a family history of early onset, genetically associated breast cancer, cannot face her day without an unreasonably firm conviction that vitamins, diet, and exercise, in a precise combination, can certainly prevent cancer. The third world farmer, whose children's lives can fall hostage to drought, or to market fluctuations on commodity prices, needs to believe that making proper offering to the small statue in his one-room, thatched-roofed house, will safeguard the lives of his cherished offspring. The mystery of life and death itself, embedded as it appears to be in irrevocable and eternal loss, may be enough to block wonder from the hearts of millions of people, who are rocked into comfort by repeated, ritualized, propitiation of personified cosmic powers, and by the fantasy that they have an unctuous personal relationship with the creative forces inside of quasars, dark energy, and atoms. What little they have been taught to know, from one single book, they hold in their mind free of alternatives.

The human need for security based upon certainty has frequently been commented on. Freud felt that people are primarily emotional, driven by fear of the large, dominant male in the family, and also by fear of death and obliteration. In *The Future of an Illusion* (Chapter Thirty-One), Freud described humankind seeking safety from our worst fears by reversing them, denying them, creating an imaginary father who protects his children from time and death. During the course of the twentieth century, Freud's provocative insights were supplanted by psychology that emphasized thought patterns rather than emotion. An eminent spokesperson for cognitive psychology is Daniel Kahneman, whose career touched many major universities and culminated with a long tenure at Princeton, and who shared the Nobel Prize in Economics for his descriptions of how our cognitions influence macro-economics. In his legacy volume, *Thinking, Fast and Slow*, Kahneman has portrayed our minds as rich and varied, but nevertheless dominated by the primitive predators' need to reach rapid conclusions to insure food and safety in unpredictable environments. According to Kahneman, "Sustaining doubt is harder work than sliding into certainty…we are pattern seekers, believers in a coherent world…" (This echoes Mlodinow, in Chapter Eight.) Kahneman describes our minds under the pressure of "…a vigilance we have inherited from our ancestors…" to make sense of a chancy world by elaborating narrative fallacies by way of our "…sense-making machinery…tidy, predictable, coherent…." He sees humans as having "…unlimited ability to ignore our ignorance…" and controlled by "…the anxiety we experience when we fully acknowledge uncertainty." The "demand for illusory certainty" leads us to seek "absurd faith" buttressed by "…a community of like-minded believers." The human illusion of control, and suppression of doubt, create a systematic cognitive bias of "…excessive confidence…inability to acknowledge ignorance and uncertainty…."

It is not a coincidence that this portrait of human beings under the thrall of unshakeable beliefs was written by a holocaust survivor who experienced first-hand the tenacity of ideology. Kahneman's Nobel, and his widespread influence throughout behavioral sciences, derive from the massive documentation by himself and many other researchers, that over-confidence in unsubstantiated ideas is the central force within economics, politics, religion, investing…almost anything. It is this over-confidence that we encounter in our daily conversations with would-be cancer-gurus and Middle-East fixers. Kahneman's colleague, Harvard cognitive psychologist Daniel Gilbert, has written in *Stumbling on Happiness*, "We cannot do without reality and we cannot do without illusion…." A protective grillwork of security operations locks our minds inside familiar and socially reinforced conclusions. Social discourse is only transiently information sharing, and is predominantly used to anchor the comfort of feeling: "I know!"

Maybe wonder gains its high sheen only for the solitary, preliterate hunter alone on the seal-hunting ice, or for the post-modern intellectual, or for the gifted artist, or for the rigorous scientist. Or maybe wonder is scattered in archipelagos of people who differ from their neighbors only by having been touched in some way—by genes, art, science, good luck, or grim fate—by wonder's wand. Some people preserve wonder in a reservation in remote corners of their psyche: island vacations, mountain hikes, evening reading. Other people may be waylaid by wonder and imploded by it through the medium of a book, a college course, or an earnest friend. For a lucky few, wonder is an organization of character. For most people most of the time, preoccupied with tasks, wonder is pulsatile, like the Cepheid variable stars whose luminosity is periodic, and who shine down on us rhythmically to pervade us subliminally.

Wonder is a multi-dimensional human capacity. Because we are nurtured and developed in relationships, in silence and in speech, wonder capitalizes on all of our relational modes to the "other." Our capacity for imprinting, for deep, sustained, enduring devotion, is echoed in wonder's power. Our primate, information-gleaning adaptation, speaks to the utility of wonder. And the universe of emergent laws, in which circles and cascades of information-driven regularities interact to generate more intricate and self-complexifying regularities, to create our woven world of life, is also inside us. The biophysical, psychological substrate of wonder within us utilizes the same capacities as the universe itself.

In moments of wonder, we vibrate to the world's complexity in such a way that our own ancient forces and creativities reverberate. Our trillions-fold cosmos of human cells are connected up and awakened while receiving the world's import. Billions of years of DNA evolution shudders in memory and recognition.

SECTION III

Crowds of Atoms in Communion

Chapter Twenty-Nine

Atoms Strategically Placed

The next three Sections discuss the wonder of life. There is an intrinsic problem to this exposition. To the extent that the narrative merely recapitulates biology—biochemistry, bioenergetics, molecular biology, cell biology—it will become massively technical, and will require a lifetime, or many lifetimes, of study. But to the extent that the exposition of life ignores the scientific facts and amplifies merely the rhetoric of astonishment and amazement, it will not capture the essence of wonder, which arises from the stirring impact of life's full complexity and dimensions. To capture the wonder of life through language, I believe, it will be necessary to explain detailed science well, to break apart the compartments of sciences like physics, chemistry and biology in order to observe processes as they move across dimensional and temporal stages, and to create a portrait of life that is simultaneously cosmically *panoramic*, while it also twinkles in the zone of the sub-atomic.

We will be set back on our heels with wonder when we can experience life with the scope of the expanding universe, and also with the tweezers of the world's smallest and most elusive particle/wave photons and electrons; when we can understand life depends upon both universal law, and upon accidental, one-time historical paths. The greatest impact that the wonder of life can have upon us comes from its hugeness, embedded in cosmic law, coupled to its detail, that fades into minuteness and invisibilities, all the while resting upon layers and integrations wrapped around each other for millions of years upon billions of years, an evolving process of cooperation, destruction, re-emergence, and amplification, the fitting of law, energy, and material into communicating, high-speed interactions of unimaginable and irreducible complexity.

The wonder inside us is the wonder of life in self-recognition. If you put effort into understanding what life is, you will feel wonder.

The starting place is to understand that we are collections of atoms strategically placed.

This statement in turn requires understanding what atoms are, how they are directed, and how they become strategically connected into vital systems.

Atoms can be understood as the building blocks of all matter, or alternatively, as a concept helpful in understanding how the world is put together. If we consider all our ideas to be social constructions that are subject to change when new information arises, even so, atoms remain important. Our understanding of the universe rests heavily on the idea of atoms, which forms the basis of our ability to understand all of chemistry and biology. We understand matter, whether bricks or living eyes, to be collections of building blocks, atoms. Atoms are not the only way that we think about substance, because atoms, in turn, have sub-components like electrons, neutrons and protons, which are smaller, or subatomic. Understanding the world as a collection of atoms is an essential insight that explains many features of life, though understanding the world from the standpoint of its subatomic components yields further explanations, descriptions, and accurate predictions about other aspects of reality, which will become the focus of Section IV. There are even sub-sub-atomic theories that consider, for example, how protons are formed before they combine with electrons to form atoms.

I am suggesting that wonder will be catalyzed if we think of all life, trees and birds and grass, but most importantly ourselves, as collections of atoms placed as they are for a strategic reason.

The concept of "atom" is a gateway between those realities we can contact, and those realities that disrupt our conventional thoughts and perceptions and open us to wonder. Atoms share the world we know, and atoms measure and participate in worlds that we enter like Marco Polo.

Atoms are congealed units of cosmic energy that form building blocks of matter. The relationships inside atoms, between the sub-atomic components like protons and electrons, and the relationships between whole atoms, are governed by the few great consitutional laws of the universe. These cosmic governors include the "strong force," which dwells only inside the atomic nucleus, where the protons and neutrons stay, and holds together the center of the atom.

The law of electromagnetism rules the realm of atomic combinations, in which whole atoms connect to form molecules. Electromagnetism is the king of chemistry. The congealed units of cosmic energy, the atoms, stay in position due to the laws in the court of electromagnetism.

So far, we are looking at the world as energy, congealed, forming matter, like electrons and protons, governed by laws, like the "strong force," made into effective building blocks, called atoms, which hold together due to another realm of laws, called electromagnetism. This reconstruction of the scientific worldview operates like children's blocks that stack up over and above each other. Take off the last big one, and the others can be seen serially nested inside. But the choice of *where* the atoms are placed derives from an entirely different realm. Electromagnetism holds them, but something else *tells* them where to go. The *placement* of atoms in our bodies, or in trees and grass, is not at all like another large block stacked over smaller ones.

The biological laws governing exactly where each atom stands and holds hands with its neighbors are not few and fixed. They are not like the grand old master laws of physics such as electromagnetism. These selection laws are the laws of life, and exist only where life exists. Life guides atoms to take up positions that favor more successful life. Its logic is circular, choosing itself from among its own best already existent forms. Life chooses life. We feel this momentum in our hearts and minds every moment. We love our life. Our attachment to existence is the psychological analogue of the biological survival force.

Life is a force that strategically organizes and places aggregates of atoms, and that induces coherence among them. Life is a social director of atoms.

Chapter Thirty

Capturing the Universe's Governors

L ife's laws are utilitarian, pragmatic, and situational, varying with location and circumstance, and not "hard wired" like the laws of physics. While physics' laws, like electromagnetism, seem coterminous with the universe, intrinsic to its original information state, the laws of life do not seem to preexist life, but emerge as part of it.

Life and its selection laws evolve out of the universe as the universe evolves life.

While the structure of atoms rests on a powerful, simple constitution, life is more like a cascade of inventions that invent new inventions. Built upon the constitution, and never in violation of it, and possessing its own master law of natural selection, life nevertheless is flexible, changeable, creative and proliferative of its very own Robert's Rules of Order.

But two problems are immediately recognizable in this description: selection and origins.

First, if life chooses better-adapted life from among its already existent forms, where did those pre-existing forms come from? The answer to this first problem is: from variations within the spectrum of already existing lives.

Life always harbors a vast pool of variations. Life's selections are never perfect, and never fully eliminate the variations and options that already pre-exist in the great pool. From the array of different life forms, some survive and proliferate optimally, while others survive less profusely, only covertly, on the margins. Any proliferative advantage of any one particular life form is always situational, temporary, and varied according to time and place. Life is strategic order that exists within the greater context of disorder, variation, and the need to shift strategies as environmental situations change.

Since selection is always imperfect, some unselected variations endure in the pool of life, potentially available for future selections. Life has the quality of a salad bar from which time and history choose carrots and lettuce on one day, but broccoli and cheddar bits on another day.

Whether we look at it biologically or psychologically, we can see that life is dynamic, tentative, and problem solving. We feel this need for vigilant problem solving in our heart and minds. Life is a challenge, and we are always working to rectify something. We adapt psychologically as we adapt biologically. We choose life from among options and variations, only to have to re-choose a moment later. Life is array overcoming disarray.

Life is a force that imposes order by making atomic selections that are successful in perpetuating life, and life is a force that never finishes its work. Life imposes order partially, and leaves partial order and disorder as bank accounts from which it can later draw, to make new arrangements which can once again renew order. Advantage and adaptation are always temporary, and soon in need of renewal. Life is like an artist painting with a thin red brush today, who will paint with a thick blue palate knife tomorrow.

The essence of life, however, is not just evolution through natural selection, but evolution through selection of the placement of minute atoms which form the molecules whose interactions are life. Life is selection of order and detail at the level of atomic location. The refined sifting of locations for building-block atoms changes the molecules which change the cells which change life. When a plant, animal, or bacterium evolves or changes, it does so because something has shifted in its structure or function, and that shift, in turn, is based on a rearrangement of some atoms within its molecules. Evolution through natural selection could not occur if atoms could not be rearranged.

The second problem with "life choosing life" is the problem of origins. If life chooses life, where did the "life that chooses" come from? If life exists only in the context of life, how did it ever get started? This echoes the problem we saw in Section I when we considered mere existence. Where or how did the supposed Big Bang come from? (Chapters Five and Thirteen) The origin of the universe is unknown, and the dignified debate about it equally upholds alternative hypotheses.

Similarly the origin of life is unknown. Life expresses the ordering forces in the universe in the realms we call chemistry and physics; and life expresses the ordering forces that emerge over time by evolution through natural selection of those vital organizations of atoms that are best adapted to survive. The survivors carry their successful strategies with them, until they are replaced by other survivors, better adapted to emerging contexts. But the *origin* of this process remains debatable. Life as we encounter it is law selected slowly in the history of place. Earth is a place.

Some people, like Jacques Monod and Stephen Jay Gould, who we discussed in Section I, see the origin of life as mere chance within circumstances that were themselves random, like the chaos of asteroid impacts, or like plate tectonic movements churning continents and oceans. Scientists like Gould and Monod believe that life might well never have existed if certain conditions had been different.

Other people, like Christian De Duve, (Chapter Eight) see life as an inevitable outcome of chemistry in a proper location like Earth, given a long enough time. Life is seen as a destiny forced on chemistry by natural law. In this view, life on Earth may indeed be based on situational good fortune, like Earth's exact rotational distance from the sun; but life on many planets in many galaxies is destined to arise out of a positive bottleneck of universal laws. De Duve describes life as "combinatorial imperatives moving downhill through constraints." He uses the forceful term, "imperatives."

Origin-of-life research has been a vast scientific enterprise for decades, and remains unfinished. Over my own lifetime, this exciting field of inquiry has grown and matured, but remains inconclusive. It would be wonderful to know the origin of life, if there ever were in fact a single time or place as the origin, but that knowledge is neither available, nor essential for deep insight into our own existence. For now at least, and maybe forever, life can be best understood by recognizing what it already is. Its origins three or four billion years ago on Earth may not even have been its origin, since it is also possible that life began elsewhere and came to Earth, (for example, on asteroid impacts), but its existence now has been deeply studied. What we are, as we know it here and now, paints a high definition, three-dimensional canvas.

Today we have the opportunity to be stunned, puzzled, and edified into wonder by our own lives as we participate in and orchestrate our becoming. Within the billions of trillions of miles of the universe that we see, its cold emptiness, we have captured and activated the universe's laws and governors, we juggle its particles, and surf its electromagnetic waves, and hold up the torch of consciousness to which we have given birth from our collated oscillations.

Every human being contains and expresses properties of the universe. Only recently have we been able to understand this truism at a factual level that unites libraries of evidence into data-rich narratives. Because we live here, now, today, we have been handed exhilarating, wonder-provoking knowledge that vaults us beyond any previous culture's gnosis.

We who inhabit the scientifically informed, intellectually open-minded, politically free, and economically providing twenty-first century, have a chance to know who and what we are in a manner that is compelling, transformative, uplifting, and hair raising, "hrstaroma." There is nowhere you can look, nothing you can touch, where, for a moment at least, as you pursue the daily ardors of adaptation and survival, suddenly the curtain can't be lifted.

As far back as civilization has left us records, even the wisest minds could not penetrate secrets that are now easily known by us. Just as we are masters of motion and can drive cars, without being mechanically brilliant, merely by good fortune of our birth in this era, so we are blessed with the knowledge of our culture and our historical moment. The most mediocre among us can easily answer some questions that stumped Einstein and Darwin. We are commoners in a rarely beneficent, opulent, generous kingdom of knowledge. We know some percent of what was once the mystery of life, and we can walk with face uplifted towards the other percent.

Arising out of invisible origins, law and our personal histories have built up for a brief moment, grain by grain, our world-mirroring, wonder-drenched life.

Now we can see that we are partly agglomerations of congealed energy that coheres as atoms, which hold formations together through electromagnetic law, and which were placed side by side by the laws of history, location, historical circumstance, and adaptive evolution. But we are much more than that too.

Chapter Thirty-One

Atomic Mazes, Canals, and Circuitry

L ife means atoms strategically placed, and life also means energy moving through atomic channels that are specifically structured to direct, contain, and regulate energy. Life is structure, formed by atomic placement, and life is the flow of energy within those structures.

We are mazes and roadmaps, through which the energy of the universe travels according to the traffic rules that our inner atomic and molecular streets require. Life is the creation and organization of energy-holding structures, made of atoms, where the flow is given pause, and stands for a moment, or circles slowly. Life is a unique organization of very high specification. It took billions of years to arrange the details of how our three dimensional atomic structure could be built and in turn could harness the energy flowing from sunlight and captured by photosynthetic bacteria, algae, and plants.

Life is a particular constellation of matter that creates a particular network of channels for the movement of energy.

Because it has been built over such a long time, and because it crosses the gaps between generations, life can also be described as a *memory* about how energy has been directed and utilized, and is the transmission of that memory. Because life occurs within a membrane, or a skin, and is a site, life can also be described as a watershed into which the cosmos pours energy, to irrigate matter. The matter forms canals down which the energy moves and by which it is guided and positioned. We call this flow-directing circuitry, "life." (The use of energy by life, bioenergetics, is the focus of Section IV).

Life is always interconnected and transitional. It receives, holds, and passes on both its matter-structures of atoms, and its energy flux. We are transformers and relays. We are intrinsically communicative. We receive energy, hold it, then pass it on in changed form; and we find and keep atoms for a while, and then pass them onward, too. Life consists of changing subunits within changing wholes built on an inner grid, "the information state." Life always exchanges matter and energy within the larger context. Life exists only within the matter and energy donating universe, with which life ceaselessly communicates. Life is not distinct from, but is an appendage of the universe.

Our intuitions that we are enlivened by sunny days and that we are connected to the cosmos are not mystical but factual. We are each singular letters in a message that we feel but cannot translate alone without our entire cosmic context. As we gradually unravel this translation, as we study the Rosetta stone of the molecule and the cell, the message of wonder emerges in front of us. Our energy and our atoms are borrowed from the universe; our laws derive from the initial information state and its elaborations and derivatives; and our message has as its context the complete edition of the cosmos.

So far we have considered life to be both a collection of strategically placed atoms, and also a guided flow of energy through the webs and networks created by those atoms. The concepts of "matter" and "energy" are classical concepts that derive from Aristotle, and they remain useful, though they are somewhat misleading. Both the matter of our living tissues, (those strategically placed atoms), and the energy that flows among and enlivens them, are twin channels that have a common source. Both are emergent from the early universe.

Chapter Thirty-Two

A Recent Appearance of the Original Burst

In Section I, we discussed the Big Bang, and the very strong evidence that the world around us is expanding out of an ancient, hot, compression that began some thirteen billion seven hundred million years ago, more or less. We also discussed uncertainty about the uniqueness of the Big Bang in time and space. When we consider "matter" and "energy," I would like to abandon the term, "Big Bang," a coinage which in fact began as a sarcastic jest, and refer instead to "the early universe."

In the early universe, primitive components of matter, even simpler and smaller than atoms, minute densities of energy congealed, out of the hot compressed, but cooling and expanding world, to form things like protons and electrons (See Chapter Forty-Eight). All matter condensed out of early and original energy. Only about three hundred eighty thousand years later (can we say, "soon later" in such a billions-fold scenario?) hydrogen and some helium atoms (as opposed to merely subatomic components) formed out of the more primitive plasma fog of subatomic parts. Einstein's special relativity has demonstrated that matter and energy are interconvertable, $E=MC^2$. The Manhattan Project proved matter-to-energy conversion indelibly when matter, in an atomic bomb, was converted into explosive energy. Our understanding of the early universe now makes it obvious that matter, atoms, are converted energy from the opening womb of the early universe. Matter and energy appear to be very different, but they are twins from the same birthing event. At a deep level they are forms of each other. In his biography of Einstein, Walter Isaacson quotes the fact that the energy contained in the mass of one raisin is enough to run New York City for a day.

What then, is meant by "energy" and where does our own "energy" come from?

At the surface level, our personal energy comes from our food, all of which derives, directly or indirectly from sunlight, captured by plants. The sunlight itself is a product of hydrogen atoms fusing inside the sun.

The way in which the sun works may be more complicated than we currently understand. Here, we will focus upon the important general principles, and not haggle over newly emerging scientific details, such as whether the matter in the sun is best referred to as "hydrogen," or as a "plasma" of hydrogen's component protons and electrons. For clarity, we will continue to refer to the substance of the sun as hydrogen (rather than plasma) and its process as hydrogen fusion (rather than proton-fusion). Stars like the sun may clump not only from the gravity of their own matter but from the gravity of dark matter, too.

The giant, hot mass of the sun creates so much gravity inside itself that the hydrogen atoms are fused into each other, forming helium. Two hydrogen nuclei are smooshed to form one helium nucleus. Excess energy left over from the fusion flies out as electromagnetic wave/particles, called photons, which we conventionally call sunlight, which radiate through space to Earth. (There are other atomic fusion reactions too, creating new elements in the sun). We are all products of sunlight.

Since the energy of sunlight is released from its billions of years of storage inside hydrogen atoms, we are also directly lit by the energy of the early universe that got stored in and then released from hydrogen atoms. Our personal energy is a recent appearance of the ancient and original burst, after its numerous reincarnations in atoms, suns, and plants. We are a local packet of recently re-molded energy that was born in the early universe. The hot, small, early universe poured out its energy, which flowed downhill in a cascade of plasma, atoms, suns, hydrogen fusion, sunlight, plants, food, us.

This is our autobiography. We have smashed through conceptual limits of individual sciences to see the whole thing in one sentence. This is our story, the bigger story. We are animated by the energy of the origin of the universe.

Our energy also can be said to derive directly from the early universe, not only because it comes to us from hydrogen atoms, but also due to gravity. Gravity appears to be one of the universe's master laws, a fundamental clause in the cosmic constitution, like electromagnetism. Gravity appears to be essential and elemental. It cannot be reduced to an ancestor. It does not seem to depend on some other property. Gravity is the force that arises from the bending of space that, through compression, and finally fusion of solar hydrogen, releases the energy that becomes us. (The strong nuclear force also plays a role). Gravity is the philanthropist who disperses light and life to us from the safety deposit box of hydrogen. Gravity is also the conductor who brings together the dusty atoms into great balls of fire like the sun, or into cannon balls, like Earth. We are will-o-the-wisps on the stage built by gravity. We are in gravity's own universe, and its donations are our sustenance. Our bodies every day run on a recently repackaged little collection of original cosmic energy that has been handed over to us gratis by hydrogen atoms and by gravity dealing it out from the sun.

This is why we can be described as transformers of cosmic energy. The original cosmic energy, that became protons and electrons, that became hydrogen atoms, that coalesced in the sun, that gravity compressed and fused, released energy that flew down to Earth and has become us. Now is the historical era when human beings can understand themselves as transformations of cosmic energy, without side-glances to apology or metaphor. The common facts of high school science now validate our identity in a new light. Our "cosmic origins" are not a "New Age" idea. We have a factual way to understand our relationship to the world. It is not just a matter of scientific information in textbooks, but a matter of a new realization, a new identity, for us. The growth of empirical knowledge has created a new insight into our origins, our nature, and our self-understanding.

Rather than children of sunlight merely, we are slow forms of original emergence. *What began in the beginning continues in us.* The sun itself was only a space station on the flight of original atoms and of original energy stored within the atoms. We ourselves are a brief pause on the long, long flight. We are one local reformatting of the ancient and original expansive flow.

We are eddies in the stream that flows from the first day of the wellspring.

Chapter Thirty-Three

A Selection and Rearrangement of the Original Moment

Though it took its own sweet time to get here, the original energy is what we harvest from our cheese and toast. We are one unfolding arrangement of the cosmic flux. We are the fallen angels from the hydrogen bombs in the sun, themselves only one more in a chain of ongoing transformations of original matter and energy.

Similarly, our atoms are strategically arranged by the universe. Our grains are gifts of the world. We are pure energy, and we are energy in the form of matter. The origins continue to play out within our personal dust. Out of the soil of the cosmos, its grainy atomic matter and its electromagnetic waves, we are pumpkins in the garden. We are a selection, rearrangement, and compression of the origin.

About fifty percent of a human being's atoms are hydrogen, and some of the hydrogen atoms in us may be over thirteen billion years old. Each one of us should celebrate his or her thirteen billionth birthday, or half-birthday. Some of our hydrogen-half may be as old as the sun's own inner hydrogen. We are carriers of the first formations, congealed from the original conversion of energy into atomic matter. We surf on the beachhead of the first cooling down, the original atomic kernels and seeds. Our matter includes *the* matter. We are pregnant with the mother load. Our bodies' hydrogen atoms are located both in our water, which is hydrogen-heavy H_2O, two atoms of hydrogen for every one of oxygen, and also in most other organic molecules like proteins, DNA, etc. Many of our hydrogen atoms are original cosmic creations, founding citizens, the universe's "D.A.R." No one got here before we did. Our tissues and structures are the handicraft of the universe.

Both our energy and our matter are temporarily borrowed from the universe. We are entirely on loan.

The elements in our body other than hydrogen are relative newcomers. The carbon, oxygen, nitrogen, sulfur and phosphorous of which most of our other half consists, along with many other lesser components, were made in stars or in super novae, from more complex nuclear fusion reactions secondary to gravitational compression (plus other physical processes).

Our personal *carbon* atoms are probably about six billion years old, made by now deceased grandmother suns, which preceded our solar mother, and then blew up, and spread their elements, hydrogen, carbon and the rest, out into the world, from which our sun and solar system including Earth, re-congealed. Stars are born and die, like everything else. The new atoms of new elements forged by the heat and compression in stars and supernovae are blasted out into the universe when the stars die. (There are other scenarios of stellar death, too). The extruded solar material may, under some circumstances, be re-contracted by gravity to form a new second-generation star. That was how our own sun is thought to have been born a "mere" five billion or so years ago. Our elemental carbon and oxygen atoms, made in one or more grandmother suns, though younger than our hydrogen, are still venerable grains. Their birth is unimaginably ancient. Though relative newcomers, they are older than our star, having come from one of her mothers. In terms of sunlight and energy, the sun is the womb of our biosphere, but the newest atoms in our own body are paradoxically a generation older than our mother, the sun.

Carbon is the genius inside life. Its atom, with six protons, neutrons and electrons, has a unique capacity for electromagnetic bonding, connecting itself with other atoms. It is the community organizer of the world of atoms. It creates the matrix within which life forms and gains the ability to build structures which can channel energy fluxes. That master of connection and affiliation, carbon, our six billion-year old elemental bone, is older than the Earth, older than the ball of the sun (though not older than the sun's hydrogen). Carbon is a product of stellar nuclear fusion reactions, and our own carbon, upon which we hang all our flesh, comes from our mother sun's mother.

Why is it considered metaphor, exaggeration, or intoxication to refer to our life as cosmic? The pebbles in our cement have been tumbling in the mixer of space and time since the old rotor got going.

Chapter Thirty-Four

Sentences Never Before Spoken

That our energy is ancient and original, that our atoms are ancient and original, that our carbon-based chemical skeleton was a product of a grandmother sun's alchemy, does not necessarily satisfy another aspect of our nature. We are a long-woven, unimaginably complex *organization*. We are not only *what* we are, but *who* we are. While we are direct legatees of the early universe, its energy, matter and first blossoming burst, we are also brand new, unique, one time de-novo mosaics, distinct and irreproducible patterns. We are not only our components, but our interlocking arrangements. Every mollusk, monkey and person is a singularity, a strategic particularity. We are atoms organized as never before and as never again. We are not just stuff and flow, but we are new selections and positions. We are creations from the universe's studio and we deserve a signature of the artist in our lower right-hand corner.

The two most fundamental facts that we can know about ourselves are that we are inlets of the cosmos, and that we are originals in the art gallery of the world.

It is often said that life is characterized by reproduction. This has some truth but it blurs an essential fact. Non-living things, like stars may also reproduce. Our biological form of reproduction, after the invention of sex and genetic recombination, is designedly imprecise. We always "aim off," intentionally inexact. Humans, mammals, and other complex life forms conceive children who are genetically very similar to but definitely not the same as we are. Variation is guaranteed, because every human mixes the genes of two parents. We generate offspring and transmit genes but we do not clone. We replicate genes, but then mix those of two parents, to produce innovations. We *transmit* but we do not reproduce. The planet and the universe proliferate uncountable new atomic recombinations—us.

It is as if we speak but do not echo. We all are saying something partly known and partly new. We are all born as sentences that have never before been spoken. In the beginning was the word but we are all new phrases. The universe has a constitution, an original information state, a core group of ancient simple laws but the universe is always saying something new. We are quotes in the latest edition of a newly edited text.

The biological uniqueness that we all possess is a fact mirrored in our psyche. Our individual existence is a lonely, isolating prison, and we yearn to fuse with the whole, as Einstein, and philosophers of almost all religions, have taught. In our cells, one by one, and in our compacted bodies, we are the first, last and only pattern that we are. We will never be duplicated. Each person is unique, temporary and alone. As entities, we are solemn solitudes. Our energy-guiding tunnels, and our atom-lined corridors, are very similar to and totally distinct from other living beings. We are each locked in genetic privacy that will never recur. Despite what you have read in every biology book, life is not reproduction, but the impossibility of re-production, transmission without reproduction.

Maybe we are straddling a line of both luck and opportunity. We are cosmic and we are local. We carry the world's laws, atoms, and energies, and we have personalized fingerprints and signatures. As for escape from the prison of self, as Mark Twain said about avoiding the vagaries of New England weather, "Wait a minute." Our dual nature makes us part-of and different-than. We emerge out of and look back at the universe. We can alter emphasis on either pole of our nature, its unity with, or isola-tion from, everything else, to see ourselves as cosmic and local, both. The classical mystical or religious dualism, that we need to affirm both our uniqueness and our belonging, can be made more accurate to science by adding a third pole.

Although we hold the jewels of the cosmos in our hearts, and although our presence is brief in time and infinitesimal in space, we feel closest to, and the greatest wonder about our complexity.

Chapter Thirty-Five

Cohering in Unison

Complexity, irreducible complexity, historically-compounded, incalculable amounts of order and precision: our wonder at ourselves and everyone around us is based upon the emergence of so much information in such a brief, small locus as ourselves, when the black expanses of the universe seem so blank and clueless. The appreciation of our information-density is one of the radical realizations uncovered by modern sciences.

Jacques Monod, only fifty years ago, imagined himself to be alone in an unfeeling immensity (Chapter One). Our dazzling new inheritance reveals that we contain in our selves a cosmos of something equivalent to Einstein's "supreme intelligence," not coded in one mother formula as Einstein incorrectly imagined, but in its elaborately woven coherence without the limitations of formula. There is no one simple powerful formula. There is layering, threading, and compounding of law with law with combination and permutation. For us to lift an eyelid, raise a finger, hum one line of a Grateful Dead tune, all the realms of existence that we have named—physics, chemistry, biology, cosmology, causality, history and so forth, and all the regular or occasional patterns they produce, are threaded into and produce the tapestry of that moment.

Jacques Monod's sense of existential absurdity and isolation no longer seem descriptive of how we see ourselves. Against the pressures of expansion, dissolution, and dissipation that rule the Hubble universe, for a moment we touch inside ourselves ordered internal galaxies, crowds of atoms who bow together in communion. Our atoms march not only to a different drummer, but to a very complex Mahler symphony. One of the wonders of the world is how unspeakably and immeasurably complicated we are. We are not only cosmic and unique; we are the most complicated compounds imaginable. We are octillions of atoms set in place by the slowly steeping, mingling, and interacting laws.

We are force fields that temporarily reverse the great explosion of the Hubble universe's expansion. We are compact, compressed and dense. Our octillions of atoms cohere in unison. Our own life is a galaxy of atomic *relationships*. In the touch of tiny atoms to each other, a new kind of universe emerges. This happens in every person, every mouse, every carrot, every streptococcus bacteria, every living thing. The Talmudic adage, "Whoever saves one life saves the world," is not mere sentiment.

But we face a tension to which Stephen Weinberg and Jacques Monod have pointed. Despite the informational density in ourselves and every other life, the universe seems reckless in its explosive emptiness.

The universe began with and continues in expansion. Dispersal seems to be the essence of the cosmos. The expanding universe continues to surge outward, becoming increasingly thin and empty. The universe consists of things moving apart, disorganizing, cooling, flying away from each other. This "entropy" seems to be the opposite of the processes associated with life. The universe seems to be running down towards an undifferentiated, disaggregated, chilled-out randomness and emptiness. Cold space with only a miniscule degree of heat above absolute zero is already the main feature of the universe. The universe as a whole, the explosively expanding early universe, the Hubble universe in which everything is running away from everything else, seems predicated on the principal of decay. There are billions of trillions of miles of empty space in the visible universe. Hubble himself estimated that the average density of the universe is the same as the density of one grain of sand occupying the amount of space which is in reality occupied by all of planet Earth. Every gradient, every place where one zone is information-dense and differs from another zone, seems to get leveled as the expansion into cold emptiness continues. The universe seems to be emptying itself out into distance, gradientlessness, uniform, antisocial void. The bedrock scientific law of entropy fuels the perception that the universe is "pointless," as Nobel Laureate Stephen Weinberg said, or "an unfeeling immensity," as Nobel Laureate Jacques Monod said.

Although gravity seems to pull things temporarily together into clumps of stars or galaxies, expansive forces seem dominant. (The possible role of "dark energy" in this expansion is discussed in Section VI). The entire universe is expanding, and the density of visible matter is becoming thinner. Whether all of this is due to newer hypothetical phenomena, such as dark energy, or not, the implication for us seems the same. The disordering forces of the universe are pervasive, and penetrate to the smallest level. The oxygen in our air is bouncing, jostling, and ricocheting against other oxygen molecules at the rate of ten billion collisions each second. The universe seems to be essentially dissipative and chaotic.

But there is life. The compaction and order of the atoms in the life-system, Earth biosphere, seem to move in the opposite direction from the universe itself. Life seems fundamentally connected, regulated and systematic. Every cell—and we will soon try to consider how many cells there are—is a functioning system that greatly exceeds our intellectual understanding. Every cell is a masterwork of handicraft. How can an expanding, cooling, leveling, and dissipating universe contain an apparently renegade planet upon which life is augmenting complexity, relationship, detailed and skillful atomic touch? How can we each be octillions of atoms communing in unison when everything else is disaggregating and disorganizing?

As a great biology text *Molecular Biology of the Cell* by Alberts et.al. puts it: "Cells must race against the unavoidable processes of decay which are running down hill toward greater and greater disorder." How and why did our cells join this race against entropy?

Life's apparent escape from the laws of entropy is usually explained away in classical science as an illusion. In fact, in terms of the movement of energy, "thermodynamics," and in keeping with the Second Law of Thermodynamics, which defines the law of entropy, life-forms actually dissipate more energy than they collect. In every chemical reaction that builds the cell's components of molecules from the atoms of the cosmos, energy is partly stored in the chemical bonds, which hold the atoms together as molecules, but some energy is always lost, wasted, released and dissipated in the process. Accompanying creation, there is always a simultaneous increase in energy dispersal. In this sense, life does not race against entropy but actually furthers it. As we build and maintain our bodies, we use and cast off more resources than we store up or preserve. Life in this viewpoint temporarily appears to increase order when, in fact, over the long run, it accelerates entropic dispersal.

The energy trapping, channeling, and transforming processes of life are partly inefficient, wasteful, and in service of entropic randomness and decay. Locally, life seems to hold energy and expand order but, taken as a whole, each cell, animal, plant, and our entire biosphere, spreads the solar heat out into the void. The spreading of heat occurs from our body heat loss, or from our fuel burning. The sum total of life's energy equation furthers entropy. We look neat but we are adding to the mess. We participate as cogs in the cosmic engine that is helping the sun to waste away, use itself up, deplete its energy, and fade out. It is not just the spoiled, SUV driving citizens of American suburbs who are wasteful. Planet Earth's life facilitates the reduction of the gradient of the sun's heat in a cold world, and participates in the movement of everything towards the big, empty chill. As we breathe every day, as we lose body heat every day, as we eat and turn food into lost heat, we dance to entropy's tune. Life is extravagant of energy.

Chapter Thirty-Six

Airy Clouds of Yellow Butterflies

The common scientific viewpoints are to see entropy and life as opposites and antagonists, or as master and servant. In these constructions, the expansion of the pointless, cooling universe is facilitated by its tool and slave, life, which is a mere gradient-reducing physics mechanism. Our "purpose" is to increase heat loss. Seen in this reductionist way, we are not purposeless, but our purpose has no human dimension. We are puppets on the stage of cosmology, physics, and the process of expansive leveling. According to this way of thinking, "entropy" means the reduction of any concentration, smoothing out of everything from something into nothing. In this viewpoint, the heat loss generated by life's metabolism ultimately serves entropy, that is, reduces the gradient of heat from its concentrated form in the Sun, or on the Earth, and disperses it out first into the atmosphere, then into the universe, so that life, as complicated as it is, is really just a mechanism in service of basic physics. Reality is dissipation. Life is a a gear. Now we see the viewpoints of Weinberg and Monod.

There is another way to place entropy and life within the same frame. Entropy is the expansive engine that drives cosmic energy, and life rides on and harnesses entropy. Life cannot contradict the Second Law of Thermodynamics; nothing can, anymore than life, or anything else, can ignore gravity. As the original decompression and expansion of the world drive energy and matter outward, life rises up from the force. Life swirls up from entropy's outburst, like a cloud rising from convection currents of heat on a hot summer evening. The biosphere grows green and expands like a balloon filled by the pressure of entropy's wind passing underneath. As the universe spends away its energy, it buys life, at least here on Earth. The essential story is not that life spends energy, but that energy expenditure by the sun, by hydrogen fusion, by cosmic law, fuels life. There is a cost, an ultimate and final cost. Like a mother who spends some of her life force to carry and birth a child, our universe has, at high cost, birthed Earth life. As the expanding cosmic energies speed past, they twirl the molecules of planet Earth into a dance we call life.

The irreducible complexity involved in the ordering of atomic grains in the vast and precise array of life, requires and harnesses the cosmic dissipative explosion. They are not opposites but *interlocked continuities*. The powerful emergence of life is fuelled by the great original event with its enduring, extravagant, dissipation of energy. Life is a product of an explosion. Life is a child born out of energy flowing past. The universe is expanding and fading away, and has exhaled its entropic force into life. As one set of events dies off, another set of events are propelled into birth. On the cosmic scale, and on the generational scale, new life springs up from the winds of decay. The expansion of space may actually be a self-augmenting force. This is one theory about "dark energy" and accelerating cosmic expansion (see Chapter Seventy-Seven, Section III).

Loss and gain guide the world, death and birth, entropy and life, disorder and order, the Second Law of Thermodynamics and the original information state of the universe. It takes two hands to build something.

Who can say whether the universe has a purpose? We *can* say that order, order compounded, deep ordering laws accompany decay and dissolution outward. The energy of expansion, dissipation, and cooling is used to rotate the wheel of life, as water rushing down an old New England millrace drives the wheel that makes flour for bread. The gravitationally driven downhill flow of water transforms grain into meal. Entropic energy flowing past planet Earth rotates the wheel of life.

As the powerful tributaries of the Amazon River drain off the water of South America, washing everything down towards the cold Atlantic, airy clouds of yellow butterflies rise up above the leveling currents.

The galaxy of atoms touching atoms inside the universe of life flutter above the great river of entropy.

Chapter Thirty-Seven

Vast and Complicated Beyond Our Capacity

We have been considering the wonder of life from several angles: strategic ordering of vast arrays of atoms, intricately guided channels of energy, use of the original cosmic matter and energy, formatting of trillions of unique, singular lives, and harnessing entropic dissipation to generate order and complexity.

To further capture our modern wonder at life, we need to stop and muse for a moment about large numbers. When we read about life or the universe, we typically encounter huge digits. For example, one light year is nearly six trillion miles, yet astronomy tells us about thousands, or millions, or billions of light years, meaning there are billions of trillions of miles above us in the night sky. Similarly, in Chapter Three, we talked about sextillions of molecules of "ATP" made in our bodies each second. What do numbers of this magnitude really tell us?

On the one hand, we have to employ numbers to make useful calculations. On the other hand, numerization contributes to a false sense of comprehension by reducing dimensions to tallies. We imagine we understand something that we simply have counted. Science uses numbers to contain or to compute. But wonder is a form of contemplative expansion that is camouflaged by computational containment. The very *dimensions* of life are part of what shatter our routine perceptions, and expand us toward the wonderful. Life can be counted or at least estimated, but it cannot so easily be grasped or comprehended. Life spills over our jars of numbers into amounts we can't domesticate without destroying the magnification it contains. Life is big and a lot. If we use them with discretion, numbers can be used to capture the wonder of life without reducing ourselves to mere tabulations. We need to clearly differentiate what we really grasp and comprehend from what we can merely manipulate and quantify. Then the enumeration within life's processes will reveal its Titanism.

Let's start with Carl Sagan's statement in *Cosmos*, that the human body contains 10^{28} atoms. If the single clearest statement about life is that it is one strategic arrangement of atoms, Sagan's number swings the door of life wide open into an astronomer's world-view. How did he come up with that number, ten octillion atoms? We would have to poke an unbelievable quantity of BBs into position in order to become ourselves. This number does not seem arrangeable. How can such a titanic number of tiny bits become organized into an entity?

First, let's look at the problem of counting. The average lifespan, let us say, is eighty years (now that we have running water in our homes, jog, and have antibiotics), of about three hundred and fifty days, (rounded off) of twenty four hours, of sixty minutes of sixty seconds, for a total of approximately two billion, five hundred million seconds. If we were born in the act of calculating, lived long, never slept, and crammed two counts in per second, we could never top a count of five billion. If we took birth as enumerating research assistants in the late Dr. Sagan's Cornell lab, and pursued our count past retirement till death, we would have left octillions of our own personal atoms uncounted! Our count would still be more than nine hundred and ninety-five billions away from a single trillion. We would not have depleted even the first of our many leftover octillions. Certainly the number "10^{28} atoms in our body" is a guestimate. Nobody can actually count that much. The magnitude of time, the beats of our inner baton, our intuitive human sense of life, is far too small to "get it" when it comes to our own real size. There is nothing in our lifespan that prepares us to apprehend magnitudes like "octillions."

The statement that we each contain about ten octillion atoms is a generalization, because we all differ in size, and vary slightly in genetic composition. The number is based on various extrapolations, such as the number of cells in one human body, the number of molecules per cell, and the number of atoms per molecule, each one of which would be uncountable, and at best a ballpark estimate.

There is another way to arrive at an estimate of the number of atoms that are strategically placed inside a person. Every atom has an "atomic weight," based on the number of protons and neutrons in its nucleus. (There can be slightly different weights, or isotopes, of the same element). Chemists have long established a relationship between the weight of a single individual atom, and of a group of them, in which the same weight in *grams* of a substance as its atomic weight for a single atom, will contain 6^{23}, or six hundred sextillion atoms of that substance. (This huge unit is known as a "mole"). From an estimate of a person's weight, let's say about 70 kilograms (70,000 grams or 154 pounds), if about half of a person's atoms are hydrogen (of atomic weight = 1) then only about ten percent, or seven grams by mass, is hydrogen (because it is so light compared to other atoms), and seven times six hundred sextillion atoms (or seven "moles") comes to four octillion hydrogen atoms. Since that is about half our atoms, we have about eight octillion atoms. Even if we say that the estimates of octillions are way off, say, a thousand-fold over-estimates that have been created by someone trying too hard to argue a point, we could then say we have about septillions of atoms in our bodies. There are two important points to these otherwise silly, speculative, order-of-magnitude statements.

Numbers that seem at first to be "scientific" and "meaningful" are often falsely mentally organizing and essentially meaningless. If you are told you contain ten octillion atoms, and later hear that you were subjected to exaggerated over-interpretation, and really contain only ten septillion atoms, how does that alter your understanding? Numbers of this magnitude serve utilitarian functions, but they are incomprehensible. We can compute with such numbers, but we cannot grasp them. Even altering them a thousand-fold does not change our emotional, conceptual insight into the nature of our life. We cannot conceptualize what we state. We know facts that we can't integrate into meanings.

Something deeper is at stake here. If we contain septillions or octillions of atoms, and life is the strategic placement of atoms, and if our intuitive capacities are proportional to our lifespan, which contains some paltry billions of seconds, all of this helps to explain why we find it so difficult to define and understand life, and ourselves.

The organized selection of and placement of so many parts, to create a functional whole, is both the basis of who we are, and is something we cannot grasp. We are vast and complicated beyond our capacity to think or imagine. There may be some ways in which we will never be able to understand who we are…Einstein's "limit."

Modern biology has created an encounter in which our living, waking reality is shattered and overturned. We are entering domains of knowledge with details so minute and with numbers so large that our waking-world proportions fall away like discarded debris burning up in space behind a Cape Canaveral rocket. We are learning not only new facts but new ways of relating. We are shocking to ourselves. We are bound and sealed together out of a manyness whose meaning must lie beyond the words and numbers we use to point towards and signify it. Towering forces of *order* have been generated by the stream of entropy that is running in the universe beneath us.

Chapter Thirty-Eight

Layered, Woven, Ancient and Embedded

I f we say the dimensions of the universe are about fourteen billion light years wide (does it matter to you that the number "fourteen billion light years" is an approximation for "thirteen billion, seven hundred million light years?) and a light year is about six trillion miles, so that the universe is something like 84 sextillion miles across, does that help you to emotionally understand the universe? I am not implying that scientific accuracy should be abandoned. The point is that, when we switch from quantification to insight, new types of information become instructive. And, in any case, we ultimately have to recognize that the incomprehensibly huge numbers with which we try to scientize and relate to the universe are nevertheless too small, or barely adequate, to relate to the numerical dimensions of our own bodies. The statement that we are dimensionally cosmic, far from being an overstatement or hyperbole, is actually understating the case.

Our atoms numerically outstrip stars and light years.

We are dimensionally trans-cosmic. The specific, precise, and functional array of atoms within us requires guiding information that took billions of years to evolve, that organized septillions or octillions of units, and that eludes our conceptual mechanisms. Our patterns of thought-dimensionality, which evolved to help us hunt, mate, and survive, cannot hold the facts of our internal order inside a telling, evocative framework. *We cannot narrate a simultaneously accurate and meaningful story about our informational complexity*. Our framework of facts produces statements whose implications are incomprehensible. Science speaks with numbers whose trajectories escape from the gravity of our emotions.

Although we are not trans-enumerable, we are trans-knowable. We are not just ordered, but we are layered, woven, ancient and embedded. We are ordered with a magnitude that journeys beyond anything we can conceive or feel. The dimensional magnitude of the order inside of us is a source of wonder to which we will return repeatedly in the upcoming chapters.

We cannot elevate ourselves so high that we can state a purpose for life and for the universe, but it seems both arrogant and counter-intuitive to reach the conclusion that life is purposeless. Nor can we imagine a big old Titan shoving into place with a skillful finger each one of our octillion (or so) atoms. No space probe has located the mighty abacus that this feat would have necessitated. And so far, our musings have considered only one body on our widely inhabited, tree lined, grassy, shark infested, mouse-over-run Earth of seven-billion-plus humans. Many, many atoms are in place in our cell-based, green and crawling biosphere. Even our grand numbering schemes fall down helplessly in front of our new, scientifically derived perceptions. We do not have the capacity to fully appreciate our layered, woven world. There are trillions of lives, each of which contain an array of septillions (or so) atoms strategically placed.

When you look around you at the world—the honey locust trees in bloom, the men with pot bellies and grey pony tails, your child or wife—and you can hold for a minute in your mind's eye not merely that they are built out of octillions of atoms, but also that the atoms are integrated, cohering, and well-placed, you will have changed the tallies of science into the realizations that transform your sense of reality.

The eye that sees wonder also sees "original information," "emergent law," "more is different," "irreducible complexity," "woven order," "numerical gigantism," "cosmic dimensions," and "compounded development."

If you keep these scientifically derived phrases in mind, the horizon-less world will never fade into safe and boring assumptions. No theology or anti-theology will seduce you. You will stand there with your hat off, wondering.

Within an obscure corner of the cold and empty universe, this intensity of reliable atomic connections has coalesced. The rules of which the cosmos consists, "the initial information state," and then the weaving and complexifying of those rules together over huge spans of time, has created energy-harnessing, atomic assemblages that have elaborate, internal road rules. The confederation of atoms of which we consist outnumbers the stars and cooperates better than spouses. The cosmos of ordered atoms within life ascends like a rainbow over the black emptiness of expansion, dissipation, and space. We are not very big but we are the product of cosmic energy, matter, law, history and weaving. To measure the multi-dimensional complexity of which we are, the best yardstick is the sense of wonder.

Chapter Thirty-Nine

A Dynamic Concatenation

We will now move our sense of wonder past atoms, into the multiple domains of units by which we can understand, but also be aghast at our own becoming. Consider cells.

The atomic communities of life are held in corrals, cells. As some ancient American civilizations were built on two seeds, corn and beans, the universe has two kernels, galaxies and cells. A widely quoted statement in biology books is that a human being is an organization of about one hundred trillion cells.

You will quickly recognize at this point that no one has counted them, and that this is a statement intended to inform, but which is a better metaphor of magnitude than a fact. The human body contains an overwhelmingly huge number of cells to orchestrate into an entity, each one of which has an overwhelmingly large number of atoms cooperating as molecules to orchestrate into a vital cell. Although the number of cells is magnitudes smaller than the number of atoms that we have just been considering, it is still incomprehensibly large.

We are so proliferative that if we were told that in fact we contain only ten trillion, and not one hundred trillion cells, the loss of ninety trillion of our purported cells would not worry us. We can conceptually jettison ninety trillion cells without feeling diminished. With or without them, we cannot really believe or understand our estimated census. Your nervous system alone is thought to consist of between one to ten trillion cells which form an interconnecting network of an estimated one quadrillion "synapses," or connections between the cells. Each cell can make connections to many others. (In Chapter Nine, an historically earlier estimate of one hundred trillion connections was mentioned.)

Each one of us is an integrated kingdom thousands of times more populous than the seven billion human inhabitants of planet Earth. Since those humans are an often embattled, nuclear armed, resource-depleting mess, the obvious question is, who rules our trillions of cells? How are we able to hang together? What is the governing system that integrates so many cells and so many atoms?

Inside life, levels of information intersect, understand each other, and cooperate. A lattice of laws connects physics zones of atomic connections, to chemical zones of molecular compounds, to life zones of cells. For example, the electromagnetic force, one of the universe's most basic statements, which tells protons, electrons, and atoms so much of what they are allowed to do or not allowed to do, is responsible for positive charges attracting negative charges. Because of this, atoms, with their various valences, combine through lawful informatics in the zones we call physics and chemistry, to make the exact combinations of molecules, including the large, unique carbon-based macromolecules of life. Thus, basic laws of the universe, like electromagnetism, lie inside the emergent zone of biology and life. To some extent, our bodies operate on the same laws as planets and stars. However, that is only our foundational, or basement level of operations.

The molecules in cells vary from the simple and tiny, like two oxygen atoms, which form an oxygen molecule, to gigantic molecules, like DNA, which contains many billions of atoms. The basic laws like electromagnetism appear to be subservient to the more complex biological, cellular laws, but it is also the other way around. The old royal family of fundamental rulers of the inorganic world, like gravity and electromagnetism, create the fences, the constitution, outside of which life's legislators cannot step.

But the laws of life escalate rapidly into massively interwoven communications of their own. The inorganic bedrock laws lie inside and guide life, and life takes these laws and with them weaves new and surprising patterns. This is similar to the way in which the few simple constitutional laws undergird and lie inside the massive pages of state or town level legislation added year after year, which never violate the federal constitution, but which are much more specific and applicable to local problems.

Our human languages provide another metaphor, or may actually follow the same rules as the languages of life. From twenty-six letters, libraries emerge (see Chapter Twelve). Almost limitless information can be compounded out of twenty-six signals. Similarly, the molecules within cells are built out of sequential signals that act like words, in DNA, RNA, and protein. However, life adds an even more complexifying dimension. Molecules, once they are built, interact with each other, adding, limiting, or regulating each other's activities. Our entire civilization is not necessarily more complex nor more intra-communicative than any one of our trillions of cells.

Linear information sequence in molecules, compounded to interactional mutual regulation, leads to libraries of communicative capacity in every one of our trillions of cells.

We are products of *governance and communication* inside the cells of life. A dynamic concatenation of statements made by the universe guides atoms into molecules, and the molecules then make limitations and directions, which guide the formation and activities of other molecules.

Core physics laws format the universe and limit its arrangements like a constitution; then, new laws weave new laws, molecules direct and control other molecules, which give feedback to and direct and control the molecules that direct and control them. Order and information compound and complexify. The recognition that information contained within sequences can self-augment through self-interaction is one of the repeating themes to the wonder of life.

Chapter Forty

Billions of Quadrillions of Atomic Positionings

We get a glimpse of life's dimensions and complexities when we consider that we are compounds of uncountably large numbers of atoms, strategically placed into molecules of uncountable variety, held in functional communication-systems of cells numbering approximately one hundred trillion, which cohere and relate as an entity, the living person. We are more than a hive, more than a web. We are more numerous in our parts and more intricate in our details than any galaxy or anything else. Because there is nothing to which we can liken ourselves, our complexity is trans-metaphorical. The only comparison to ourselves is us. "Us" includes elephants and mosquitoes. All of life, not just people, contains dimensions and complexities that escape out of and hover free from our ordinary comprehension. To engage the wonder of life, we need to develop new thoughts and feelings. Alongside our new information, we need new ways of holding it.

We have already gazed into the numerical incomprehensibility of our atoms. Let's consider the molecules into which the atoms are compounded and which form the working components of cells. Some of the molecules of which we consist are small and other molecules already partake of the dimensionality and complexity of life itself, even though the molecule is only a subcomponent of a cell. Today our highest profile, publicly recognizable molecule is DNA. DNA can be estimated to contain approximately hundreds of billions of atoms. Every cell has DNA, so our personal collection of DNAs contain something of the magnitude of ten septillion atoms (that is, hundreds of billions in trillions of cells) each atom in his or her own place. This incomprehensible numerical gigantism is nevertheless a magnitudinal *underestimate* of our atomic complexity on two accounts.

First, our body currently has its trillions of cells, but many cells are made and remade recurrently. Some cells, like the neurons in our brain, may be made once, to last a lifetime. In a centenarian, many of her neurons may be one hundred years old. But many other tissues, like skin, gut and blood, make, break down, and remake cells rapidly. Billions of our cells die every hour. A neutrophil, which is merely one of the many types of white blood cells, may be made one hundred billion times a day. Our blood cells, like fashion models, are always of the latest edition. If we cut our finger, we expect our skin to create new, intact tissue in a matter of days, because we know our skin to be a renewable resource. One expert estimate is that the single zygote cell, the fusion of paternal sperm and maternal egg from which we all began, has to divide ten quadrillion times on average for us to live out the typical human lifespan.

This would add an enormous magnitude to our estimate of atomic placements inside all of our DNA across our lifetime. We create many more cells than we consist of at any one moment in time. So our DNA would require hundreds of billions of atoms correctly placed in ten quadrillion cells. Billions of quadrillions of atomic positionings would be involved in the creation of only our cellular DNA, leaving aside the other thousands of complex molecules of which we consist. Each of these atoms is guided into place both by information about where it should go, and by energy-requiring skillful processes that *implement* the information.

Secondly, in considering our atomic complexity, as we have seen, life has layers of laws guiding it. When two atoms of oxygen join together to form molecular oxygen, they do so following the commandments of the basic information state of the universe, the electromagnetic force, in interaction with more specific laws, which we label, "chemistry," that are dependent upon the configuration of protons and electrons in the oxygen molecule. When the approximately one hundred billion atoms inside of the cell's DNA take their places, their locations are initially designed by the laws in the realms we refer to as "physics" and "chemistry," but they are also controlled, guided, and placed in groups by laws unique to life itself. These are the selection laws of biological adaptation, that, over billions of years, collected and saved sequences of atoms bonded into molecules, which in turn formed the alphabetic code of information that characterizes DNA.

DNA contains the memory of life that survived. The memory is coded into the sequences of which DNA consists. These sequences create dictations, rules of life, inside the organism. The sequences code biological laws, that exist over and beyond the physics and chemistry realm laws that dictate atomic placements. So any one atom is placed where it is, due to laws that come from the realms that we call physics, chemistry and biology, all active at once. This is what is meant by my terms, "information compounding" and "woven complexity."

The number of atomic placements in one DNA molecule is many billions; the number of atomic placements in all the DNA in all of our cells at any moment creates "incomprehensible numerical gigantism;" the number gets expanded by cellular turnover with its need for new, more DNA; and finally, the information guiding this process is as old as the universe and as new as today, a compounded weave of history, memory and transmission—all based upon the proper detailed placements of atoms. Each one of us has DNA that follows cosmic constitutional law, "the original information state," such as electromagnetism; and each one of us has absolutely unique and new individual DNA sequences, our cellular fingerprint, by which we can be accurately identified and arrested for our crimes.

Galaxies seem small and simple to those of us who have contemplated life with twenty-first century knowledge. Human society will soon grow weary of people who think they know and understand. The best way to comprehend life is by examination through patient, well informed, wide-angled wonder.

Chapter Forty-One

Reverence for What Is

Before Section III ends, let us look at some of its implications. Wonder, as we have already seen, is not always cozy. It can up-end platitudes, conventions, and assumptions. In exchange, it brings accuracy, uplift, and wisdom.

There is a drought-proof well of wonder available to those people who, within the frustrations, dissatisfactions, and losses that inevitably accompany life, are able to keep in mind the realities about life that have been revealed to us over the past few generations, and which continue to unfurl around us. Undistinguished moments of the day will nevertheless gleam with "contented dazzlement," and personal crises may be leavened by recognition of "new magic in a dusty world." We are privileged to understand the wonder of life even as we continue to struggle for our personal safety, survival and comfort.

For the wonder-sparkling, wonder-abeyed sensibility to come alive in daily life, the first step is to identify, not objectify. Science books or planetariums can convey amazement but not wonder. The wonder of life points to us. Every person, me, you, is a cosmic concentrate. We hold the highest pedigree, purebred from the cosmos, and we express evolving opportunity within constraints. Awareness of our ancestry and lineage adds impact and clarity to our self-perception. As amazing as are photographs of spiral galaxies taken by the orbiting Hubble Space Telescope, our own inner worlds have equal magnitude, multitudes, and primeval origins. Twentieth century biology has given new meaning to the old adage that heaven is within you. The gyroscope of cosmic law whirls within us. Human life is much more radiant and improbable than a star. We share in a star's nature, and we share in an increasingly more elaborate nature as well. Stars are much bigger and brighter, but we are orders-of-magnitude more complex.

There is a subtle, important difference between living in awareness of the wonder of our life, and the cloying self-adoration of the so-called "anthropic principles." The anthropic principles are ideas that innocently started when the innovative physicist, Brandon Carter, asked the counter-factual question: what would the world be like if any one, or many, of the mathematical constants within physical laws had different numerical values? For example, what if, in Einstein's discovery of Special Relativity, $E=MC^2$, C, the speed of light, were a little faster, or slower, or had been cubed rather than squared? Using the same counter-factual supposition throughout the realms of physics and cosmology, many scientists have mathematically demonstrated that minor changes to apparently arbitrary numerical values of nature's constants, would disrupt the universe and would make life impossible. This line of thought is called "counter factual" because it changes the fact and then examines the implications of those changes. The central discovery of all these suppositions is that our life depends upon the universe being exactly as it is.

Why is the speed of light 186,000 miles per second, and not 185,000? Or my favorite counter-factual puzzle: Why indeed does "plus" attract "minus" and not another "plus"? (Freeman Dyson, Richard Feynman and their colleagues invented "Quantum Electrodynamics" which describes and predicts electromagnetism—plus attracting minus—as a product of quantum events. This elegant theory, famous for the diagrams Feynman used to explain it, cannot tell us *why* things are as they are, and has some flaws that lead most scientists to view it as brilliant but incomplete).

Because the cosmos fits into itself, seems sewn together and integrated, and because it is the stage on which life has appeared, and because seemingly minor and arbitrary disruptions of it at numerous junctures would have precluded life, some people like to believe that the universe was *intended* for life, designed for life. There are several differing formulations of anthropic principles which stress more or less emphatically the idea of intention and design, and they echo the political fantasy of "intelligent design."

Although I experience the wonder of life, and want to convey why by writing this book, I do not want to be misunderstood as promoting ideas that purport to know the long term goal of our universe. I have not gotten close enough to the universe's embryonic start or to its senescent demise to issue proclamations about the intentions or the destiny of the exploding billion-galactic cosmos.

The ultimate counter-factual argument, as old as human thought, is simply, why is there anything at all, even an intelligent designer? If the fact that the world supports life proves that the world was motivated by inscrutable beneficence, then it is equally inscrutable that the world supports the intelligent designer. Where did he or she come from? Who designed the designer? If the designer did not need a designer, then neither does our world.

Why is there such a lackadaisical progression, such an unbelievably procrastinating oscillation, of building suns and blowing them apart, building dinosaurs and then, like a peevish child, smashing them all back into clay again? Why not just design a person of the preferred ethnic group and the proper religion, and stand him up at a podium in place, sermonizing on the basis of one absolutely correct book right from the cosmic start? The belief that the universe was designed intentionally to make life on Earth has to contend with the questions: then why make 100 billion galaxies, trillions of stars, uncountable planets; or why take fourteen billion years to do what was intended on day one? Why make the hideous and swallowing destructions of black holes and quasars? Why give tyrannosaurs a hundred million year long part in the play, and then edit out the entire scene? (In Section VI we will consider the ways that science answers the conundrum: why is there something and not nothing? In Section I, Chapter Thirteen, we already puzzled over quantum emergence).

Life fits the universe as it is, not because the universe was designed to create life over stumbling, chaotic, and erratic billions of years, but because life is a derivative of, an expression of, a manifestation of a potential contained within the universe. That potential is an inexplicable source of wonder, as is the existence of the universe itself. Positing either or both to an intelligent designer, or to intentional anthropic principles, only explains the inexplicable through another inexplicable, and a fantasy one at that. Why explain away our greatest moments of wonder by reducing them to impaired logic and credulous fantasy?

It is an admittedly stirring insight that our existence depends upon the exact mathematical relationships of lights and particles and laws, as we find them operating in the universe today. This certainly triggers our sense of wonder. It does not prove that we know *why* the universe exists and what its *goals* are. After all we, like all species, will certainly disappear. (This was discussed in Chapter Twenty-Seven and will be discussed further in Section VI).

Why is it surprising that life as it is requires the universe as it is? Had you been supposing that life was incompatible with the universe as it is? Because life emerged from and expresses things as they are, it fits.

The theory of multiverses, as we discussed in Chapter Thirteen, is sometimes invoked to explain the "fit" or "readiness" of the universe for life, for, if there are infinite universes, most of them may be "chaotic" or "dead" while one in a haystack—ours—may be the rare, chance, life-tailored one. According to this view, life "fits" inside our universe not due to some intention directing the whole process, but because there are billions, quintillions, in fact infinite entire universes that are "dead." Our universe would be the one rare duck egg out of the supremely vast collections that did indeed have the correct physics constants (light=186,000 m.p.s., etc.) for life to occur. We would be the improbable event that became manifest only after infinite throws of dice. As this theory seems so invented and unsupported by, say, photos from the Hubble Space Telescope, I find it intriguing, but wonder-impoverished. Let's return to the universe we can actually contact. Let's return to a world where creative speculation needs to take into account some evidence.

There is no way for a reasonable human being to assign a goal or an intention to the hundred billion Milky Way sized "island universes" that wheel for sextillions of miles in the known cosmos around us. The universe birthed life, at least on Earth, because the universe has some internal coherence. It follows a constitution, scientific laws, an initial information state, the origins, purposes, and intentions of which are no more and no less obscure than mere existence instead of non-existence.

Life fits into the universe because life is a *culled formatting of interactional potentials that the universe makes possible.*

Out of almost unlimited potential, life is that which was selected over long times and many circumstances. Positing an incomprehensible, invisible, "Other" does nothing to explain the incomprehensible "other" that is palpably present, and that we actually *encounter every second within and around us.*

When the dross and dysfunctional is sieved out through the grating of eons, life is the gold nugget that has been kept. The universe created life by stirring and permitting hoards of atoms over eons to interact endlessly, and by preserving those composites that were *viable, rugged* and *self-perpetuating.* We have been created by the laws, materials, and energies inside us.

The "anthropic principles" elaborate fundamental confusion between observation and inference. They confuse observation about the intricate, interlocking, slowly proliferative world, with inference about its' intentions, as if "it" has "motives" and "goals." They are sensitively attuned to the world's wonder, but reduce thoughts to projections about the world's psyche, which they re-construct with the narrowness of their parochial self-adoration. (Section VI will discuss "projection.") They believe in "motives" and "intentions" which they invent. They hold it credible that the maker of quasars, that destroy billions of suns, enjoys their singing and likes the man in the shiny tie who appears on their T.V. station. It takes a swollen ego to claim you are privy to the full plan of the billion-galaxied cosmos.

In the shortest and darkest speculations, our cosmos is expected to last hundreds of billions of years before it either expands out into smooth nothingness or starts its cycle of contraction, yet "anthropic" prognosticators have appointed themselves as interpreters, spokespersons and pinnacle life forms of the incomprehensibly vast spaces and long times before and after. As we have already seen, each species lasts a limited time, measured in millions or many million of years, and our Earth and even our Sun are temporary densities in a changing world. It stretches not only credibility, but dignity, to appoint oneself regent to this vastly more wonderful cosmos.

Self-appointed narcissism is not religion. It is pseudo-religion to cage the universe in security-seeking fatuous explanations. Anthropic thinking usually has political agendas.

There is a trend to simulate a dialogue between "science" and "religion" to show that you are open minded, or to proclaim that science and religion share core goals and values. As we saw in Chapter One, Einstein testified that science and religion were twins, but then went on to rule out all existing religions from his personal definition of it! The problem is actually that "religion" refers to a wild and clashing diversity of types.

In a *New York Times* editorial, October 18, 2011, two self-defined Evangelical Christians, Karl W. Giberson and Randall J. Stephens, both on the faculty of Eastern Nazarene College, attacked politicized and anti-scientific fundamentalism for its "...simplistic theology, cultural isolationism, and stubborn anti-intellectualism." They describe religious "...rejection of knowledge" as "...abandonment of the intellectual heritage of the Protestant Reformation..." and they conclude that there is a chasm between their own Christian faith and its political imitation, which they see as: "...an occasion to embrace discredited, ridiculous, and even dangerous ideas." In contrast to the pseudo-religious, (actually anti-religious) "overconfidence" of fundamentalism, they underline *humility* as the insignia of authentic worship.

A worshipful attitude is a true measure of wonder: reverence for what is. Worship contains *a receptive and expectant deference*. We see around us ageless interactions of unoriginated laws. This vision exceeds any coarse, familiar name, and any unctuous, one-sided, fantasy relationship. *Felt* religion does not address the cosmos on a first name basis nor appropriate its black, shining wonder for brief, local, political agendas.

The holy resonances echoing within our quadrillion-fold DNA atoms can be heard by anyone who listens to inner feelings we all contain about the wonder of our own existence. We worship in wonder because we feel so incomprehensibly present and blessed.

Authentic worship is a community celebrating together their gratitude and wonder.

Chapter Forty-Two

Where Information Ends

Instead of the counterfactual, let us stand on the factual. We do not know why there is something, not nothing, nor why the speed of light has the value it has, but we know it does. The best label for the scientifically compatible world-view is the "Cosmo-thropic." The universe makes itself. The universe unfolds. The cosmos in its integrated entirety weaves causality of many forms with the paths of history, into increasingly complex and emergent creative newness. At the core of it all is not a human-like motive, but inexplicable, incomprehensible wonder. It is through the feeling of wonder that we can hold all that we think and feel about life's existence, with maximum information, open-mindedness, and emotional depth. *Wonder is the epicenter of devotion, reverence and humility that we call "religion."*

I am not implying that dry, fact-hunting science drawn from one era of history gives us final answers. At the same time that we embrace evidence, reason, and discovery, we can also lift our minds higher by abolishing "data- entrenchment." Science can be used in service of either wisdom or timidity. I do not prefer the blinders of scientific reductionism any more than the pseudo-explanations of pseudo-theologies. Since our information is so evolving and incomplete, why use it to enforce false conclusions? At some point our thoughts always step out beyond the current collection of facts.

Do I have to reach out my hand for the latest issue of a scientific journal as my cremation is about to commence? Do I have to die in stolid refusal to surmise "…wild surmise"? Because I am scientific, do I have to expect the scientific world view, as I know it today, to be absolute, irrefutable, and final?

The wondering mind asks: what is most probable? Where can I stand with a mix of intellectual integrity and expectant unknowing? Fervent believers and fervent disbelievers share the strident tone that reveals their desperation to prove something by attitude rather than by discovery. I prefer truth to fabrication, and where information ends, I prefer wonder.

Flexible, credible faith is as much an opening as are new discoveries. Rigid belief and disbelief are both oxcarts in an age of cosmic questions. I know that I cohere to some degree, for a period of time, because the universe coheres to some degree, for a period of time, and I know that I do not fully know how, and I know that nothing moves and thrills me more than the enterprise of exploring this emergent communion.

We can use the Stonehenge of facts standing upright around us to align our sight. Science is a microscope and a telescope, not a lens cap that blocks the passage of light. We can use science to guide the light, to narrate credible, convincing, informed and interlocking probability statements, luminous at both ends. We can also look out beyond the known world to the farthest edge of evidence, reason, probability, hypotheses, and bemused possibility. We can continuously parse the difference between two mental states: we want to avoid pandering to ignorance, infantile security-operations, and counter-factual fables. But, because our time here is limited, we need to gaze out to the green flash after the sunset, to construe possibilities that glow somewhere beyond current fact. We don't want to shut down our thoughts either due to credulous belief or bitter disbelief. We want to keep trying to understand. Wonder is a recipe that mixes both *science* and *reverential religious feeling*.

Wonder thrives on continuities. Nothing in the universe is stationary; the hard balls of atoms or electrons are themselves oscillatory and fluid, so there is also no final form for our minds, which gain in dimension and feel full of wonder when we identify with, rather than point at the universe, when we surrender claims to final truth, when we revere evidence, and when we step forward modestly and boldly to surmise when evidence ends. There we find ourselves in a darkness that glitters with our own magnitude and detail. We are the first era of people to whom "know thyself" combines cosmology and molecular biology.

This is what we can say:

Inside the violent, entropic, dispersing universe, things have been put carefully into place, over a long period of time, into very large arrays, in very small spaces, according to an elaborate scheme, with a result, if not an intention, that new properties of energy and matter emerge, as life, as mind, as biosphere, as me.

Life reflects magnitudes and layers of laws, of information, that cannot be calculated and that defy our conceptual capacities even if we could enumerate them. This mixture of panorama and pointillist-ism runs on cosmic energy flow. How? More specifically, how can the vast, wild forces of the cosmos—the forces that speed galaxies billions of trillions of miles apart; the forces that career in thermonuclear solar explosions—guide with pinpoint precision the innumerable array of teensy atoms in my cells? Our question is about a bull in a china shop. It is the problem of the Big Bang in a cell. How can the explosive and entropic cosmos flow knowledgeably and delicately enough to touch and place octillions of atoms in lumens and lanes that twinkle with love and poetry inside of our embodied minds?

Life is beyond imagination. It is the handicraft of oscillation, precision, testing and time. Its methods cannot be guessed. They need to be explored with lenses of patience and wonder. If we are a flow of energy, exactly what does that mean? Once our atoms have been strategically placed, how do we regulate, control, and use, the cosmic energy that flows through us?

Section IV

Energy of the Biocosmos

Chapter Forty-Three

Light Sizzling Between Atoms

Somewhere millions of miles away across frozen black space a furnace is roaring. The furnace is bigger than anything in my world, bigger than anything I can imagine. In its seething core, what I call the "world" is being transmuted. In the furnace, elemental protons from the nuclei of hydrogen atoms are being melted and smashed. Energy, the force, explodes outward, moving as fast as motion can move. Energy moving on nothing, mass-less force scatters at maximum speed in every direction. Tens of millions of miles away a large rock intercepts a ray of that energy, and on the rock, a green layer captures, holds, and tosses around the speeding stream of light. The layer eats and incorporates the energy of that light. Using that energy, the thin layer builds very small bridges among tiny things. The architecture of atoms grows, a civilization, a universe within the space of a thin green leaf. In this universe of inter-touching atoms, thousands of shapes and forms touch, weave, bend and dance.

From the furnace of destruction, our sun sends light, which green leaves use to create the chemistry of life, and we are made. (The sun can also communicate with the Earth through other energies than sunlight. In solar flares, coronal mass ejections, and other events, electrons, gamma rays, X-rays, etc. can be ejected from the sun and shot towards Earth.)

Many scientists study the details of these remarkable processes. It is hard, very hard, to really understand that *we are the light of the universe sizzling between atoms* in innumerable, nearly perfect, *suspended arrays.*

We are atoms strategically placed, and we are energy magnetizing arrays of matter into men and women. It is the energy of light, which means the energy that the sun has released from hydrogen atoms, which means the energy of the early universe, that holds together the atoms whose alignments we are. We are the products of light engaging and entraining matter. Whatever its "intentions," "goals," or "motives," the cosmos has made life. We are biocosmos, energy that is alive.

In Section III, we introduced life as matter and energy arranged by information. The focus of Section III was atomic arrangement. In Chapters Thirty-One and Thirty-Two we discussed how the atomic arrangements form circuitry to direct the flow of energy. We saw life as *matter irrigated by light,* which is released from the energy that the universe stored in hydrogen. Our own lives are "one local re-formatting of the ancient and original expansive flow of energy." Now we will observe in detail what it means that we consist not only of matter, but of cosmic energy.

To understand our own energy, we will need to reach across terms and specific sciences, to stretch our minds from electrons and the chemical bonds they make between atoms, across photosynthesis and respiration, down into the *transportation of energy by electrons* that is the secret and essence of our lives. Once again we will encounter within ourselves all the complexities of the universe. Our vision of life will be both high-definition and on a very big screen. We will need to keep in mind complex causality, as discussed in Section I, and the multiple domains and dimensions of knowledge that our current cultural moment has awakened out of the previous slumber of uninformed belief.

By keeping in mind the reach of time involved in the evolution of life, we get a glimmer of our own magnitude. By reversing scientific objectification, and by attempting to point to ourselves as manifestation of the phenomena revealed by science, we can dwell in astounding self-knowledge. By pushing ourselves towards the limits of understanding, we can break through into that stretching, outreaching, full apprehending that feels so wonderful. When we understand ourselves as energy, we will want a painting as detailed as a Frans Hals and as big as a Michelangelo ceiling. We will probe exactly how the whole cosmos moves inside of us.

We may hope for nothing less than a transformation of consciousness by understanding who we are.

We have already looked at matter in the following way: through its atoms, the matter of life is coterminous with the universe, of which it is a temporary selection and collection. (Chapter Thirty-Three.) This atomic subset of the whole is highly selective, like an elite college in respect to the general population. What it contains, and in what proportion, makes the atomic material of life a master chef's mix. The material of life is a cake-mix of the universe, but it is very different in its proportions from the universe itself. We saw in Section III that, for example, the universe is predominantly hydrogen, with a good dollop of helium, and all other elements are mere spicing; but life is only fifty percent hydrogen, and rich in carbon, oxygen, nitrogen, sulfur and phosphorous, with essentially no helium. Life is a distilled perfume of the universe after a sifting and sieving.

Similar to our matter, the energy by which we build and maintain ourselves is also a local form of the universe, and, while it is obtained from a carefully culled process, it is also different in its very form. The matter, the atoms of life, are selected, but the energy of life is both *selected* and *transformed*. The energy that is us is wave/particles, electromagnetism, electron motion, chemical bonds, in informatically guided streams, whirl-pools, eddies: change.

Chapter Forty-Four

Photosynthesis is the Cervix

Pre-scientific mythology is replete with images of changelings, such as gods who become human, or evil forces who can cunningly morph. These myths are cherished because of deep truths they evoke. Through photosynthesis, carried out by chlorophyll and its relatives in green cells, energy from the universe becomes bios, life. The energy of life is as much a changeling as are the mythic gods. Physical energy follows simple, pervasive rules, the constitution or basic information state of the universe; but life has cascades and tapestries of laws woven among laws, complexities uniquely derivative from specific historical paths, and multitudes of interactions that communicate, regulate and propagate. Life transforms and proliferates the universe's basic energy in a manner analogous to the way that electric current can fan out through a computer with internet access to a trillion websites.

The energy from the sun that approaches the green cell is like money dispersed by the state legislature into a bank account, while flowing out of the green cells after photosynthesis we have something as complicated as a Big Ten university, with fifty thousand students studying hundreds of topics on a two thousand acre campus of buildings, professors, bike racks, libraries, and coffee cups. Cosmic energy before photosynthesis is non-specific, like money, while after photosynthesis it becomes entangled and threaded into detailed patterns inside trees, birds, and biologists. During photosynthesis, pure electromagnetic energy in the form of solar photons (released during hydrogen fusion reactions in the sun) are captured and converted into energy that can *combine*, *converse*, and *cooperate* unimaginably. The energy of life was once simple waves. When it exits after photosynthetic transformation, it is a participant in an expandingly complex creation. We can say that to make life, cosmic energy is both selected and also transformed from physics realms into chemistry and biology realms, to become the branching energy whose weave we are.

All green cells perform photosynthesis: the vast networks of ocean algae, the grasslands of Nebraska and Kenya, the forests of Pennsylvania, Brazil and Borneo, the North woods of Canada and Siberia. It is not surprising that even cities have grassy lawns, parks, and trees, whose greenery is the gateway to our own vitality. It is not surprising that ancient Hebrews venerated a tree of life, or that Indian iconography sat the Buddha under a tree. Before humans knew, we intuited, that we are products of the photosynthetic wizardry of green cells.

There are also energy sources other than solar hydrogen, sunlight, and photosynthesis. Geothermal energy is heat created by gravity compressing the material of the Earth against itself and by radioactive decay of elements within the Earth. The Earth is too small, and its gravitational compression is therefore too mild, to fuse and rupture atoms, (as the sun does) but the heat generated can be harnessed by life on an entirely different historical trajectory from our green biosphere. In the ecosystems around geothermal vents, where cracks in the deep ocean floor allow the Earth's core heat to well up, there is life in the absence of sunlight, oxygen, or photosynthesis. The bacteria and other organisms who live in these dark oceanic chambers actually take their energy from the chemical hydrogen sulfide, which they oxidize. We have only begun to come to terms with the masses of bacteria and other lives who populate the ocean around geothermal vents and who are so different than we are, and who may nevertheless contribute in some way to our world. It is also conceivable that elsewhere in the universe there could be biospheres dependent upon energy sources other than hydrogen fusion inside of stars.

Still, when we sit in a park reading a book, and pause to look around us at the leafy maples, dogs on leashes, or boys and girls playing soccer, everything we see, and everything in ourselves *by which* we see and think, is a product of the early universe, hydrogen atoms, solar hydrogen fusion, photons of sunlight, and the transformative baptism by photosynthesis. Photosynthesis is the cervix through which our waking world is born. Photosynthesis is the birthing canal that populates the surface of the Earth and builds what we call the Biosphere.

From the supposedly dead, impersonal atoms of the universe, and from its featureless flow of light energy, photosynthesis shakes and awakens the conversation among the atoms to become a combinatorial hubbub, which becomes the biomolecules utilized by our bodies. As atoms inside us communicate more and more, as our organic chemicals grow, diversify, and influence each other, they do so using the energy that photosynthesis has repackaged. Even alchemists never dreamed that sunlight could be turned into grass, oaks, turtles, politicians and pundits.

Chapter Forty-Five

Life's Two Tasks

The transformation of energy during photosynthesis rests on the ability of atoms, with their impresario, carbon, to combine. We call the joints between atoms, chemical bonds, and they hold the energy of the universe laced between atoms. Electromagnetic solar photons donate the energy that resonates as bridges scintillating between atoms so that life's network can proliferate.

A small plant, say an African violet, in a little pot on my windowsill, is capable of catching sunlight, taking electromagnetic energy out of the sky, civilizing it, controlling it, and thereby sliding it into the bonds between chemicals in living green cells. Because of this skill of plants, our own bodies can eventually be made. All green plants are our ancestors and our makers. If you love your maker, you love plants (and hydrogen, and electromagnetism). The energy that enters into chemical bonds during photosynthesis in green cells eventually becomes us.

The waves and disembodied particles of mere energy are lifted over into life. Electromagnetism is the constitution, the fundamental law, the Magna Carta, but DNA biology is analogous to Stratford-on-Avon. After photosynthesis, there is a new expressive stage in the world. A sharp turn has been made in the combinatorial complexity of the universe, out of which playwrights, plots, and people can eventually be constructed from atoms and sunlight.

We who tend to think of ourselves as "minds" or "souls" are also chemicals. The movement of our thoughts and feelings when we experience the sense of wonder, the movement of our fingers when we write or type, is an expression of cosmic energy that has been stored in between atoms as chemical bonds, which can be re-opened to release that energy again, to be properly routed to thoughts or fingers. Chemical bonds can be *structural*—they hold atoms together—and they can also be local *banks* for storing and releasing energy. We are energy that moves by making, transporting, and breaking electromagnetic connections among atoms. Everything we know, think, feel, is dependent upon the activity of making and breaking chemical bonds. We use energy to both structure arrays, and to juggle.

To understand our personal bioenergetics, we have to re-orient our perspective, to recognize within ourselves remarkably transformed sunlight that can diversify and particularize. We are atoms in place and atoms changing places and energy moving atoms around to form complex biochemicals.

When we stood in wonder at the incomprehensible octillions of atoms properly aligned to form us, (for example in Chapter Forty) when we considered phrases like "numerical gigantism," "complexity," and "trans-calcuable," we were nevertheless making underestimates of the amount of information that organizes us, because of the vibratory dynamism of our ceaselessly forming, breaking, and reforming chemical bonds as the mechanism of our energy flow. We are in constant self-creation. The dis-embodied early universe has at last been transformed by photosynthesis into *life's two simultaneous tasks*: building large complex *structures* of precise and skillful molecules; and *moving energy,* by making and breaking chemi-cal bonds, to build, rebuild and maintain life.

We are now taking the ideas of Section III, and making them more motile, dynamic, vital.

Every blade of grass is touched by the light of heaven which it turns into the sugar of life. Green photosynthesizers transform the universe, here in the corner called Planet Earth, because they are skillful traffic directors for the flow of electrons, by which we move energy into chemical bonds. Just as photosynthesis is the doorway between energy and the biosphere, electrons are the key to photosynthesis. In every thing green, even in grass, photosynthetically directed electrons are ushering light in to life. Grass is a marriage of atoms and starlight officiated by electrons. Ah, great Walt, who discerned the journey-work of the stars under our feet.

Chapter Forty-Six

Ambrosia from Sunlight

Photosynthesis means literally "making things from sunlight." During photosynthesis, energy (photons) flies across space and imports atoms into life. There is no shortage of sunlight here on Earth. The George Washington University Solar Institute says that all of the energy stored in reserves of coal, oil, and natural gas is matched in twenty days of sunshine on the Earth's surface. A National Geographic estimate tells us that the sun showers us with six thousand times more energy than our entire current energy budget.

Photosynthesis might be considered the cosmos's most charismatic skill. We can describe this transition in some detail, but can we really comprehend the implications of making our minds and bodies out of energy?

Photosynthesis requires water and carbon dioxide, along with sunlight, whose energy is used to push electrons out of the water, and take up new homes as chemical bonds. Carbon dioxide, that unsavory and colorless shade, and water, our earthly essence, are joined together to make sugar, which is the ambrosia on which all of life directly or indirectly feeds. Two simple molecules, carbon dioxide and water, are used to make a more complicated one, and this is the step where cosmic energy becomes "food." Humble molecules are used as a source of their constituent atoms and electrons, which become bound by energy into a new, life-giving molecule of sugar.

Photosynthesis varies slightly, depending on which type of green cell is involved. Some of its energy transfer is done by resonance, without electron motion, and many types of photosystems exist in many types of plants. My descriptions are designed to capture the essential principles of life.

Sugar made from carbon dioxide and water during photosynthesis is the base of the food chain for the visible biosphere. During this transmogrification, migrating electrons, energetically pushed from water, are donated into the sugar-creating kitchen inside the photosynthesizing cell. This is one of the meanings that "water is the base of life." Inside green photosynthesizing plants, electrons are removed from water and donated as sizzling bridges, holding together the chemical bonds of the new-formed sugar molecule. The water, stripped of its donated electrons, ends up as our beloved oxygen, which is breathed out by plants at the end of their tour-de-force

As long as we stand back and peer out at the world with a microscope or telescope, we fail to see our real identity. Electrons dancing across the tiny spaces between atoms create our food and our bones. Electrons in companies dance inside our minds. As part of the biosphere, dependent on the sugar and oxygen which plants make through photosynthesis, we participate in the consumption of electrons from water. Life is, among other things, skillful relocation of electrons.

Photosynthesis is considered to be about three billion years old, an edifice next to which the pyramids are the chrysalis of a few noisy insects. Photosynthesis transports electrons and in doing so makes the sugar we eat and the oxygen we breathe, and it took planet Earth more than a billion years to invent, but it has now been building the Earth-spanning biosphere for another three billion years.

If science seems too cold and logical, and if you want to raise up your emotions through adoration and worship, you might consider the sun and plants: the sun by whose brightness we awaken, whose indefatigable transit measures and parses our days, and whose light bathes, enlivens, and feeds us; and plants, who create all of our food and oxygen from which we make our bones and brains. Whether or not you are a vegetarian, you are unquestionably a breath-a-tarian.

The sun and photosynthetic plants are the brand-name company of planet Earth, and we are their knock-offs. We cannot make our own food or breath. The only real farmers are plants. Humans who tend them are only their officious footmen. We are dependent on and supplicants of the sun and greenery. We crunch and inhale the soft green electrons that have been stirred by sunlight, electrons re-arranged by the original landscape architectural team: sun and photosynthetic greenery. Our temple is the Earth, and our priests are the green cells and green plants who mediate between us and our impersonal source, the sun.

Or maybe we could spend a moment worshipping electrons, whose motions provide most of the energy for living things. Without electrons transferred by plants from water to sugar, we ourselves would not exist. It is hard to pay respects to an invisible something that hovers between existence and charge, so electron worship is an emotion for the more metaphysically religious. I don't expect to soon see temples erected for the worship of electrons, though it in fact would not be a bad idea.

Chapter Forty-Seven

The Gift

The details of photosynthesis are like jewels inlaid in the cosmic Taj Mahal.

The chlorophyll of green plants is embedded in a large team of molecules, called photosystems. (I am describing the general case that applies to all of the subtle variations of photosynthesis.) Chlorophyll, one of the greatest cosmic inventions, is a molecule that has a single, metal, magnesium atom embedded in a ring of common things: carbon, nitrogen and hydrogen. Metal is a necessary part of life! In photosynthesis, a lithe electron, loosely seated inside chlorophyll's magnesium, is kicked up by sunrays to a higher level of energy. By moving more energetically in response to a photon's knock, the chlorophyllic electron *absorbs* the photon's energy, and this is the key moment of our existence.

The energy of sunlight has increased the motion of an electron. Life has received and accepted the gift of energy from the ambassador of the ancient universe. Herman Melville, in *Moby Dick*, refers to, "…the great principle of light…" as "…the mystical cosmetic…"

Solar energy has become electron momentum inside a bio-molecule inside a cell.

The energized electron then falls, making a controlled descent through a large array of chlorophyll's companion molecules within the photosystem. The energy released through the calculated fall, like a waterfall driving a wheel, is harnessed step for step, and drives the chemical processes that make sugar from carbon dioxide and water.

The process of linking up the carbon atoms of airy carbon dioxide into large, solid sugar molecules with multiple carbon atoms firmly attached to each other, is called carbon fixation. The carbon atoms of the gas—carbon dioxide—become "fixed" into the carbon-to-carbon spine of the solid—sugar—using the energy derived from the photosynthetically excited electron. One of these steps of carbon fixation requires a facilitator enzyme, called "RBC."

RBC is the gatekeeper at the door for the main materials of all plant and animal cells in the visible biosphere. RBC, the maitre d' of green cells, is considered to be the commonest protein on Earth. This one enzyme for one chemical step in photsynthesis is the balancing point on which the gyroscope of life rotates. Yes, there are genies. We are entirely dependent upon a singular protein molecule. As we will soon see in detail, the specificity of proteins is stunning and definitional: life is the transformation of solar (cosmic) energy and matter into smart, skillful, protein chemicals. RBC is itself a product of past generations of carbon fixation, and the shop manager of its ongoing generativity.

The movement of electrons during photosynthesis to create the chemical bonds of new molecules of sugar, is guided by smart and detail-conscious molecules like RBC, who themselves are grandchildren of previous eras of photosynthetic carbon fixation. A vast array of cells, chemicals, and electron-transferring processes predate us, and were developed over slow, smart billions of years. Our great, great grandparents include chlorophyll, photo-systems, RBC, and their old fashioned knowledge about how to utilize jumping and falling electrons to harness energy. We have appeared after eons of culling and harvesting the universe's interlocking laws. All carbon fixation today is the product of carbon fixation yesterday, which made the molecules who make us.

The recipe for chlorophyll comes from earlier parental generations which created green cells in the first place. Without green cells, there would be no photosynthesis, so we are dependent upon these earlier, cellular generations. This in turn raises our old conundrum of origins. We are, ourselves, many volumes of history removed from the original simple sentences of existence, and our new drama was made possible by slowly invented electron dances. It took billions of years for cells and molecules to harness the capacities coded into electrons and to figure out how to make life out of light.

Now we can reflect on the fact that electron motility has already had two starring roles. Electrons are taken from water to become part of carbon fixation to make the solid bonds of life from gaseous carbon dioxide. And electrons in chlorophyll jump up like outfielders to catch sunlight, and then parcel out their energy to the carbon fixation team. Photosynthesis, without which large, complex life-forms like us probably could not exist, can be summarized as: waves of sunlight drive electrons out of water to take up residence as energy bonds holding together the sugar of life. Electrons are both the outfielders and the weight lifters. They catch energy and parcel it out where needed; and, in chemical bonds, such as carbon-to-carbon, they hold, store, or release energy.

Our life is the energy that is captured when electrons (in magnesium in chlorophyll) absorb sunlight, and our life is a scaffold held together by electrons forming chemical bonds. What are electrons?

Chapter Forty-Eight

Translating the Cosmic

The wakening of impersonal energy into life is possible because of the properties of electrons. To understand both photosynthesis and the chemical bonds of which we consist, we have to go deeper and understand electrons. We are who we are due to their becoming us.

To see ourselves in realistic perspective, we have to stretch our mind in one smooth gesture from the origin of energy in the expanding early universe, through the cooling of energy into atoms, and the release of atomic energy from atoms inside suns, right on down in scale to units of analysis much smaller than atoms. Electrons are subatomic choreographers of everything about us, including photosynthetic energy transformation.

At the beginning of Section III, I suggested that wonder could best be catalyzed by a discussion of life at the level of atoms. Now we are spelunking deeper and smaller.

Wonder springs up from that mental stretch across the arbitrary boundaries of scientific disciplines. One of the challenges to writing about the wonder of life is that any simple, clear, declarative sentence is in danger of being too simple. We are holding in mind a process that connects *origins* with *immediacies*, and that connects sizes light-years big with *nano-tiny* dimensions.

The division of the world into two parts, dead matter and life, is a false dichotomy. Biology and electrons are not separate, but continuities. To many of us, thinking about electrons and chemical bonds has been tainted by dry science classes of memorization and tests. Why then, is a journey into wonder plunging deeper and deeper into electrons and chemistry? Because we are turning around to understand how we ourselves came into existence, due to what has happened to electrons, atoms, and energy after zillions of interactions over billions of years. We are a selection and compounding, but we are *not essentially separate* from electrons and the universe's other "things." Electrons are not exactly in our family; we are in theirs. It is they who make the chemical bonds between our atoms. We are puppets suspended on electron joints. We are also wild deer grazing on their sweet energy.

Electrons were born out of energy early in the history of the universe, and they ferry the energy of the early universe through the narrow door of photosynthesis into the kingdom of life, mind, love and literature.

We have to shift our fundamental orientation, to recognize electrons not as little rotating balls about which we have to memorize some facts in science class, but as the messengers of the universe whose message we are. Electrons branch and flower into life, and life can then grow in awareness. When we study electrons, electrons are making it possible. Electrons are the wings inside our thoughts. We breathe because they move in cloud-probabilities. Electrons wrote this sentence and are reading it. Electrons are not "it" or "them".

Inside of every cell, inside every leaf, inside of us, lie the advancing edges of two great histories: *the history of the universe*, its hydrogen, gravity, suns and lights; and *the history of life,* cells, molecules that can trap, guide, and extract energy from electron motion. *Electrons cross between the two worlds so that the two great histories can intersect.*

Electrons are part of the primeval first expansion of energy with its attendant conversion into matter, but electrons are not exactly matter. They continue their fourteen billion year old hedge fund that invests in both matter and energy.

Electrons are much smaller than a single atom. They are too small to see. They are too small and too motile to be precisely pinned down or measured in their location and speed (the "uncertainty principle" that was discussed in Chapter Seven). They can best be discussed not as things, but as scientific ideas that we have created to help ourselves understand. We use the term "electron" to say that there is a *probable location* of a small amount of particle/wave matter/energy moving at high speed and expressing a negative electromagnetic charge. An electron is a concept used to highlight and understand elusive and invisible cosmic processes. The indeterminate, uncertain, quantum-mechanical nature of matter-energy is given a utilitarian handle with the electron concept. Not precisely locatable, fast but not precisely, electrons hold the original cosmic informational directive, "negative charge." Electrons, like the cosmos itself, evolved into existence from undifferentiated original energy. (See Chapter Thirty-Two.) When they were already hundreds of thousands of years old, as the universe cooled, they predominantly became associated with positively charged and much bigger protons, to form atoms, predominantly hydrogen and also some helium. (Free electrons in "plasma" still exist throughout vast stretches of the universe). The explanation of the existence of electrons lies in early cosmology, combined to sub-atomic particle physics. Yet these elusive entities/non-entities are determinants of life. We cannot follow Socrates' dictum, "know thyself," without discussion and obeisance towards them.

First, we need to understand how atoms join to form molecules, and how electrons make this possible. After that, we will be better able to understand how we move the cosmic energy through ourselves based on the bonding of atoms, and on the breaking of bonds between atoms, all due to electron motion. Only then can we understand how electron motion is the primal step in which photosynthesis turns sunlight into sugar. Electrons do not flow on their own, as electricity, in life systems, but move by chemically contained and regulated transfer: stepwise and constrained. (Chapter Fifty-Four)

Our chain of logic will move from electrons, to the chemical bonds they make, to a specific and remarkable use of sunlight, electrons, and atoms during photosynthesis.

The African violet on my windowsill can eat sunlight and create leaves and flowers because it knows how to play ball with electrons.

Chapter Forty-Nine

Messengers of the Universe

C hemistry is possible because electrons respond to energy.

Chemistry may be understood as *destiny*, etched into the pattern-forming law books of the universe (Chapter Eight). Combinatorialism is dependent upon the division of existence into electromagnetic plus and minus, the primordial male and female. Things combine due to this attractive duality, which is written into the primal information state.

In the early universe, positively charged and relatively heavy protons formed (allegedly made out of more basic components called "quarks"), along with smaller, more mobile, negatively charged electrons. After hundreds of thousands of years, these adolescents of the universe married; "plus" bonded to "minus" in hydrogen atoms. Hydrogen atoms, the simplest ones, can be understood (with some degree of simplification) as one positive proton electromagnetically bonded to one negative electron. In the "kitchens" of suns and supernovae, millions of years later in the wedded life of atoms, new, larger, and more complex atoms, like carbon, were formed from nuclear fusion, but still, the laws of plus attracting minus ruled. On Earth today, inside our bodies, our complex macromolecular chemistry can create molecules consisting of billions of atoms, and governing this variation and combination are electromagnetism's simple commandment: "plus attracts minus!"

The making of chemical bonds to store energy, and the breaking of bonds to release energy, generally occurs from electron motion. The "plus" charged protons are much heavier and less mobile. Generally, they stay put inside the atomic nucleus. (We will see some exceptions to this later). For "plus" and "minus" to meet and match, electrons have to do the dating. Electrons' motion is the key to where our atoms stand, how they combine, and how they guide our energy. To make a molecule, you have to remodel the very atoms of the universe. This in turn requires modifying electrons' energy levels within the atoms, and their relationships to other electrons in their own and in other atoms.

It is the elusive, multi-faceted, matter-and-energy nature of electrons that enables them to serve all chemical bonding, including carbon fixation during photosynthesis, as messengers and transformers.

Electrons are the messengers of the universe because they cross the boundaries between matter and energy, between location and speed, between two atoms. But because their motions hold atoms in molecules together, they are also the universe's glue. Our complex world and bodies are the results of electrons making bridges between atoms.

Although electrons express the uncertainty by which the universe manifests itself in the tiny quantum world, they also paradoxically express *certainty*, and are the notaries of the chemical domain where atoms combine to form molecules. Electrons behave with fidelity to the laws of electromagnetism. They absorb and release energy reliably and predictably. This is why Victor Weisskopf said (as quoted in Chapter Four) that the world has reliable shapes and forms because electrons have reliable shapes and patterns. He added: "The stability of electron wave patterns causes the same flowers to bloom every spring…"

When we say, "the living world absorbs solar energy," or "chemical bonds store and release energy," we are talking about electron action. Electrons are therefore obedient to electromagnetic law and disloyal to any one atom where they may presently temporarily reside. They may well be only a human constructed concept layered over the indeterminate quantum world, but electrons are among the most powerful explanatory concepts we have.

Electrons are a little troubling to us because they defy some of our ordinary reliable experience, and after we spend time and thought trying to understand them, we remain unsettled. Holding matter, energy, speed, location, uncertainty and change in their elusive little rocket ships, they also serve as the world's stodgy accountants who make our world reliable and predictable at least to some degree. At our level of daily perception, *electrons hold atoms in the patterns of the things of the world.*

The African violet on my windowsill guides sunlight to transform itself into electron motion. Photosynthesis is the exact location where energy waves kick and move electrons, and therefore where electron motions change atoms, which change molecules, which create chemical bonds which store and hold energy which feeds and animates leaves and love and thoughts and baseball. Yes, there is a moment, there is a place, where the universe quietly and secretly explodes into the combinatorial chemical network that branches into molecules, monkeys, mothers, and men. Like a genie out of a teapot, our biosphere pours out of an invisible, pinpoint spout.

Chapter Fifty

Montages Rooted in the Invisible

Can we really recognize ourselves as complexities suspended upon solar energy touching and connecting via electrons? Electrons *contain*, *transport*, and *release* energy by making and breaking electron-mediated chemical bonds among atoms. Our bodies, motions, and mind exist due to the chemistry of electron wave forms participating in increasingly complex processes.

In the molecules of our bodies, atoms usually join via the type of bond called "covalent." (There are a number of other types of chemical bonds also used in our cells). Covalent bonds are acts of electron sharing, by which our carbon-based chemistry creates large molecules which hold together atoms of carbon, oxygen, hydrogen, etc. You could say the atoms marry and share electrons. To the chagrin of moralists, you could also say that negatively charged electrons marry two, or more, positive proton-containing nuclei at the same time. Chemical bonding is marriage, bigamy and worse. Electron sharing, though widespread, is not so simple, because the electrons of combining atoms repel each other. The atoms are attracted to each other by electron-proton, minus-to-plus bonds, but are repelled from each other by the minus-to-minus electron charges as the two spaceships approach each other to dock. This is where electron mobility and acrobatics are necessary to form our chemically interacting bodies and world.

The electrons of covalently bonding atoms must jump around, changing their energy state, in a manner dictated by electromagnetic laws in the zones we call physics, and in a manner dictated by the numbers and states of electrons that existed in the atoms before they do their combining, that is, by the laws of chemistry. Atomic bonds always change electrons, adding energy to bounce them up into new states that facilitate atomic marriage and that reduce electron-to-electron repulsions. Chemical bonding requires two things: new *energy,* and new electron *arrangements.*

We saw in Chapter Five that Wolfgang Pauli explained this most fundamental property of the universe, that electrons will exclude each other and will occupy unique niches, private-property energy levels and spins. So when atoms bond, energy must be added, electrons must change their states, and new arrangements of protons and electrons form new chemical units: molecules. In the covalent bonds that form most of our life, the energy for all of this pops out from the broken bonds of previous molecules which we call "food." Covalent bonds within these food molecules are snapped open, and the energy of their electron states is transferred to our cells to make new molecules, out of new combinations of atoms, requiring subsequent changes in electron energy states. (Remember that some of the energy of food's chemical bonds and electron states is lost as body heat which dissipates out into the universe as entropy, as we saw in Chapters Thirty-Five and Thirty-Six).

The world inside us, of which we are an expression, is the movement of energy released by electrons (from food-molecule chemistry bonds) or absorbed and held by electrons in new energized states (in the living molecules of our cells). Our life is *energy moving with electron change in atomic bonds of chemicals*. We are cosmic energy arranged in atoms, and cosmic energy rearranging atoms into bonded molecules, and the change in the atoms results from their electrons having absorbed or released cosmic energy. Electron change underlies the entire process of our world. *Life is energy rearranging electrons* in life-giving ways.

The subatomic world is more intimate to us than science textbooks commonly expose. We are moving, breathing and thinking because electrons interact with energy, because matter and energy have a common source. In essence, we are energy moving, held, and released from electrons in atoms.

As trees are rooted in soil, we, who feel so solid and separate, are rooted in the properties of invisible and impersonal worlds. We do not "depend upon" electrons; we emerge out of their active, absorptive, releasing, reliable, disloyal natures. We are late, late consequences of electron's communion, combinatorialism, marriages and bigamies inside the constitution of matter-energy. We should not construe our sciences as peering and poking at electrons, but as opportunities to realize that we are *elaborate montages of what electrons can do*. We are their patterns. It is we who are the ephemeral collages, and our forming forces lie deep inside invisible and mercurial properties.

Covalent bonds between atoms contain many subtleties, varying among themselves in strength, angle, and likelihood. (My descriptions all contain some heuristic simplification). There are also many other types of chemical bonds, all more or less a product of electromagnetism and electron motility. In Chapter Forty-One we saw that some scientists indulge in counter-factualism about the constants of nature—why are they as they are, and what would the world be like if the facts were counter to what we live with today? Counter-factualism contains the core great question: why is the world as it is? Electrons are a good place to play this game. Why are they minus? Why are they existent? What if they ceased to follow the exclusion principle and chemical bonds did not cohere as reliably as they do? Returning to factualism, we can appreciate ourselves as the offspring of chemistry's control of energy flow, and of electron's regency within chemistry. There is no explanation for why this is so. This is a zone of wonder.

Electrons creating the labyrinths of life are also the active agents in photosynthesis, the alchemy that forms the biosphere. Photosynthesis is a story about electrons that is our own story, our real, deep "Roots."

At last we can see for energy (as we saw in Section III for matter) that our minds and bodies are estuaries of the cosmos. From its expansive oceans, through channels as narrow as a single electron, universal energy enters us. We are now capable of tracing the energy that moves our finger. It arose at the origin of time. After all, where else *could* it have come from? Yet, we are the first generations of humans who are able to think through this process. In units one electron in size, the Big Bang enters the cells of life. From within us it can be channeled to continuously new purposes.

We are embedded within, and derivatives of, cosmic becoming.

Chapter Fifty-One

Civilizing the Wilderness of Atoms and Electrons

The cosmos is mostly a cold, dark, empty place, and where matter and energy exist at all, they are mostly seething and unruly. The universe appears to be chaotic and ignorant.

Life may be understood as a form of cosmic insight. Electrons, atoms, molecules, cells, over very long time-periods, became aligned into systems so complex and so *cooperative* that they could harvest sunlight, to live on, grow on, to pass along to increasingly complex life. This insight into the nature of energy and how to utilize it required understanding about how to catch sunlight and channel electrons. Life had to "know" this.

Life can also be understood as a form of cosmic *destiny*. Given the properties of photons, electrons, atoms, and molecules, and given billions upon billions of years of bumping, colliding, and encountering all these components in events too numerous to quantify (even using the most expansive "-tillions,") it may well have been inevitable that chemistry would, like water flowing downhill, eventually coalesce into life.

We now have a scientific civilization that is capable of dissecting many components of life, like the protein RBC, or like the whole process of photosynthesis. But dissection always shortchanges one dimension, its antithesis, integration. It is one thing to juggle a ball, and it is another thing to juggle octillion balls. Processes of life always occur in larger contexts, which are in turn integrated into even larger contexts. Electrons in chlorophyll jump up to catch the energy of photons, but that can only happen in magnesium ions which only can function this way inside of intact chlorophyll molecules which can donate their electron-energy only to receptive photo-systems. How did all of these proximities and processes get aligned and integrated? Each part can do its job only when embedded in the whole system of multiple parts. But the whole system can't exist and cohere until the separate parts are already doing their job. We have a chicken-and-egg problem.

The chlorophyll-containing photosystems in green cells are integrated partly because they are *compartmentalized* in subunits called chloroplasts. Chloroplasts are walled structures inside green cells that hold many of the component parts necessary for photosynthesis in functional proximity. They are thought to have first evolved as free-living, little one-celled "bacteria," about three billion years ago! They originally lived on their own, but at some ancient time they were engulfed by larger cells. The uptake of chloroplasts by bigger cells, and their mutual cooperation, afforded protection to the chloroplasts inside, and provided "auto-tropism," that is, self-made food production, for the bigger cell that engulfed the chloroplasts. Photosystems in chloroplasts feed their host cells and live inside their protective walls. Photosynthesis is therefore a symbiosis of electrons, chlorophyll, photo-systems, chloroplasts, and cells.

Chloroplasts are little life forms that made a great invention and then modestly took their pivotal yet hidden place as the active greenery inside green cells. Chloroplasts in their modest dens are the locations where sunlight activates chlorophyll's electrons and energizes life.

Both the invention of photosynthesis, chlorophyll, and chloroplasts, and also the union of chloroplasts within larger cells that became green, are the great moments of Earth history. William the Conqueror and Napoleon are mere commas in the book of life that is now being written by the team of chloroplasts and plant cells, whose Gilbert-and-Sullivan-like cooperation is called endosymbiosis. Endosymbiosis is itself a poster boy, or poster girl, for the expansion of the role played by mutual benefit among living things.

We who dangle dependently from the ocean of air, that is made by photosynthetic chlorophyll, chloroplasts and green cells, as if we were monkeys from a vine, can also learn good manners from our microscopic nurturers. Much of the oxygen that we breathe is a product of teamwork that is more ancient than any rock that can be currently found on the surface of the Earth. Life is not only complex and creative, it is expansive, incorporative and communal. It has become a post-Darwinian commonplace to portray evolution as competition, victory, and conquest. There is some truth to this, of course, but the evolution of life also includes very important roles for communication and community. Quid-pro-quo, and mutual benefit, seem exemplified by that momentous evolutionary step, endosymbiosis, which were brought to our attention by the University of Massachusetts's great biologist, Lynn Margulis.

The earliest appearance of photosynthesis, in free-living single-cells, that would eventually become chloroplasts, long predating the cooperative endosymbiotic event, was in the range of three billion years ago. For a long time, these little bacterial photosynthesizers (who were only later to become chloroplasts) made their own food by carbon fixation from sunlight, but nothing and no one breathed in the oxygen which they cast off as an end product. It took as long as two billion more years for our oxygenated atmosphere to accumulate. Oxygen may have been initially absorbed back out of the atmosphere by iron in rocks, slowing its rate of accumulation. Nothing could depend upon oxygen when it was rare and thin.

We are discussing several life-transforming, Earth-transforming steps, each separated by gigantic stretches of time: the creation of photosynthesis by early single cells; the endosymbiotic transformation of the early cells into chloroplasts inside bigger and more complex cells; the creation not only of some oxygen as the end-product of photosynthesis, but the emergence of the oxygen blanket around Earth; and the evolutionary creation of oxygen breathers dependent on the atmospheric blanket.

Therefore, the evolution of oxygen breathers, like us, depends upon a complex weaving of many realms of scientific laws and many unique historical events: not only the universe's laws, the sun's light, the Earth's and moon's existences, water's electron-donating capacities, the origin of life, early cells, but also photosynthesizing early cells, who invented that rare genius molecule, chlorophyll, its friends the photosystems, endosymbiosis of chloroplasts into plants, and the very slow accumulation of our great blanket, host and home, the oxygenated sphere of gas around our planet. We ourselves arrived only after a very long preparation during which many features of the world had been fit into each other in a long series and in overlapping processes.

All of our energy, holding aloft all of our carbon-based macromolecules, all of our oxygen, stems from chemistry kindled by sunlight. Over a long planetary history many details have been spliced together. At the heart of it all, an electron leaps up in light. It takes chlorophyll to make a village.

Life springs from seemingly abstract laws to become beautiful and three-dimensional. Like the ballerina, an electron leaps. Like the ballet dancer, life catches and skillfully lowers the exquisite one.

What awakened person would ever tamper destructively with the algae-bearing ocean or with a tree-studded forest? Humans continue to destroy their original green lung because they continue to worship icons. The wonder of life is layered, ancient, tongue-and-groove. How long did it take the tinkering cells to locate and to place magnesium at the center of an electron-dancing carbon-based chlorophyll ring? Who can study life without having their thoughts crack open like a pistachio to reveal the kernel of wonder? We are not the imaginative inventors we like to believe ourselves to be. We are late-appearing flowers of ancient, and intricate harvests, communities, and organizations who have settled and civilized the wilderness of atoms and electrons. We are a crop from the fields of life forms. Chlorophyll and chloroplasts bequeath energy to us, and dangle us out into existence like yoyos. Out of the cosmic corridors of dreary, dead particles, teams of inventors and farmers harvested us.

We and our sciences are not the truly original creators and discoverers, but are only craftspeople in the workshop. Our own thoughts dwell within what has been created by the long, long weaving together of electromagnetism, atoms, electrons and their kin. We are the outcome, the derivative, the omelet. And the denouement in the cosmic play is the scene where magnesium, chlorophyll, photosysystems, chloroplasts and green cells teach an electron a new dance.

The combing and aligning of the wild spaces and forces of the universe, to form the furrows from which we spring, was done by legions of little geniuses. There are eons of know-how and great scientists embedded in the leaves of my African violet. We might be wise to acknowledge that the greatest inventions of all time have predated us by billions of years. Like Newton, we stand on the shoulders of giants, but in our case, the giants are too small to see.

Chapter Fifty-Two

Respiration: Regularities and Rivers Aligned

I n pursuit of the wonder of light energy within life, we have seen how electron motion enables photosynthesis. That is only half of the story. Electron transfers also create our oxygen-based respiration. Electron motion is how we humans take our own energy from the universe. To understand ourselves we need to understand respiration, the mirror image of photosynthesis. The energy that is harvested from the cosmos by the electron leap and fall during photosynthesis is then moved further downhill by more electron acrobatics during respiration based on oxygen. Respiration means: electrons falling downhill into oxygen and donating energy to cellular life. Whether you are a doctor, a truck driver, or a programmer, you earn your livelihood through respiration.

Billions of years after photosynthesis first used carbon-fixation to make sugar and, as a byproduct, began to pour out oxygen into the atmosphere, electrons and primitive cells invented a way to use that oxygen to give greater power to life. That team of inventors who tooled oxygen use for energy extraction transformed life from being strong, into being large, smart, and thoughtful. Respiration powers muscles and minds by using oxygen to maximally utilize electron motion. Respiration is the biocosmos' state-of-the-art electron energy use.

Every moment we are breaking chemical bonds, transposing electrons, and circulating energy. Just like photosynthesis by plants, our human, mammalian, animal way of getting energy out of food molecules rests upon our ability to skillfully harness the aptitudes of electrons for breaking and making bonds among specific atoms inside larger chemical molecules. But now large and intricate systems, creative and unique beyond imagining, ferry the energy from food into the molecules and motion of us. And the sine qua non for this is oxygen.

We live at the bottom of a pool of atmospheric molecular oxygen, that was created by electron-jumping, electron-stealing photosynthesis, originally by single-celled lives. Several billion years had to pass before the oxygen that was exhaled at the end of photosynthesis became deep and reliable enough for oxygen-breathing life like us to stake our lives upon it. Then and only then, the next great electron dance, respiration, could come to life.

In Section I, we considered how fundamental laws do not account for the world we see, which creates compounds of laws woven into tapestries of laws, emergent laws, phase-state laws, and path dependent histories which shape the world around us by excluding all the alternatives that were left behind. Among the complexities of regularity, we discussed top-down creation, in which new phenomena change the very things that birthed them (Chapter Ten). The oxygenation of Earth revolutionized our planet from the top down, making entirely new forms of life possible. The planet that breathed out molecular oxygen was changed irrevocably by the oxygen it had created.

The use of oxygen by living things enormously increased the efficiency of energy production that could be obtained out of the same amount of food. Like alligators in a Florida lake, we live at the bottom of a covering we can't do without. We draw in facilely, or greedily, or desperately, oxygen that is the milk from our mother plants. Atmospheric oxygen exists, and therefore we can exist in such fullness.

In the photosystems of green cells, one modest molecule, chlorophyll, contains electrons that lasso the sun and deposit its gold into chemical bonds, forming sugar, while simultaneously molecular, breathable, oxygen accretes like pearls. From the universe to the sub-atomic electron clouds, all these affinities, regularities and rivers have been aligned, and flow down into our lungs and hands, for us to breathe and eat.

The chain of becoming by which I can write and you can read this sentence includes the cosmic expansion of the early universe, the crystallization of electrons and protons and then of whole atomic matter out of the original burst of energy, the gravitational compression of hydrogen atoms into suns, the cooking of hydrogen into more complex atoms like carbon, the nuclear fusion of hydrogen to release the original energy as light, the harnessing of this energy by the photosynthesis team of chlorophyll and the photosystems, and finally the use of photosynthetically-derived oxygen within respiration to *release more energy in new ways* to enable big apes with brains.

Photosynthesis and respiration form a symmetrical planetary cycle. What one side creates, the other side consumes. We breathe in the exhalations of photosynthesis and eat its sugars; and photosynthesis grabs up our exhaled carbon dioxide and water to recycle into its oxygen and sugar. We are married to plants. We are only one half of what we really are. We, in all of our fullness, are only one side of the wheel of Earth life. Our respiration is as much an electron trick as is photosynthesis.

Chapter Fifty-Three

The Radiant Scaffold

We have been thinking about the advancing edges of the universe and of life, and their electron meeting-ground. We have also been thinking about the structure of our bodies, which channel energy flow of electrons as they make and break chemical bonds. And we have studied how photosynthesis in green cells uses the properties of electrons to capture solar energy to make our manna: sugar and oxygen. Now we are moving inside our own cells, to stand in wonder while we observe ourselves *take energy out of the universe.*

The electron journey of respiration, similarly to photosynthesis, amplifies the facts in Chapters Thirty-One and Thirty-Two, that our energy derives from the construction of skillful, intracellular channels that we build with our molecules. We concoct molecules with which to eat the cosmic heat. The manner in which we do this reveals details of unimaginable beauty and precision. No ancient fairy tale was even close to guessing the cellular refinement and cosmic participation of our electron-moving bioenergetics. What we call "ourselves" are the epiphenomena that steam up from the subatomic filigree of electronic chemistry. By moving around electrons exactly as we do, we absorb cosmic energy and speed it down pathways whose webs and weaves we feel we are.

We eat food to get our pieces and grains, hydrogen, carbon, and the rest, but we also eat food to get the energy that plants (or animals who ate the plants) have stored in the chemical bonds holding atoms together after photosynthesis. We break those bonds and recycle the energy to make our own inner maze and architecture, and use that energy for the movements of our arms and legs, for our body heat, (entropy), and for our thoughts.

When the chemical bonds in food are broken, the energy that is released cannot be simply poured out. That would be as useful as coming home from the supermarket and spilling your groceries onto the driveway.

Analogous to the manner in which you remove your grocery items from the shopping bags, carefully placing them where they belong, milk in the refrigerator and macaroni on a shelf and not visa versa, we all remove the energy of our food in careful steps. This allows for safe energy harnessing, so that no wild, unruly energy runs chaotically through us. This allows for thorough energy harnessing for maximally efficient utilization. The previous form of energy harvesting, before photosynthesis made a lot of oxygen, was skillful and careful, but was not efficient. Much of the energy of chemical bonds in food in those archaic days was not changed into useable form until oxygen-based respiration was invented.

The specific use of oxygen to get more energy out of food is probably less than two billion years old, considerably newer because it is dependent on photosynthesis (which predated it by a cool one billion years or more. The dates of those microscopic ancient events are debated). Then again, put another way, aerobic respiration is billions of years in the making. When people say that Greek Civilization is very old, or that Varanasi on the Ganges is an ancient city, or that the Bible dates from the dawn of time, you have to wonder what planet they came from. Earth's materials had uncanny, revolutionary insights into cosmology, physics, chemistry and cell biology billions of years ago. We have inherited the estate of cosmic bioenergetics, on which all of us sons and daughters are fed for life by electrons trained to serve us. Through the capacities of electron messengers, we breathe and speak. We are their inventions. We fall out of their skills and societies. If we experience the holy, it is due to them.

During our respiration, the sugar that entered the world through photosynthesis (or secondary molecular products of that sugar), are broken back down into the simple carbon compound from which they were once so lovingly crafted, carbon dioxide. The electrons that helped to hold the sugar together are released, (reversing carbon fixation, Chapter Forty-Seven) carrying their liberated power with them. These infinitesimal energy packets and waveforms, electrons, are steered down a radiant scaffold, the elucidation of which took numerous scientists decades to discern, and knowledge of which constitutes one of the true Seven Wonders of the World, on a par with galaxies, cells, or photosynthesis. The scaffold of energy release inside us is called the electron transport chain of oxidative metabolism.

Respiration reverses the carbon fixation from photosynthesis, and re-releases the carbon dioxide back again out of the sugar, and it simultaneously reverses the theft, or adoption, of electrons from water. At the end of respiration, electrons are donated back to form water again. The mutual symmetrical antithesis of photosynthesis and respiration is the great marriage, the great flow of life: simple carbon to living carbon and back; water to oxygen and hydrogen and back. Respiration is the chemistry-set mirror image of photosynthesis. Together, the two processes are us.

To appreciate the depth of wisdom, and complexity inside every breath we take, we will journey through an epic filled with familiar heroes like electromagnetic laws and electron dances, as well as a new pantheon of mobile proteins, metallocatalysis, membranes, mitochondria and even more subatomic wizardry, proton pumps.

Laws, properties, and eons, layered over and among each other from time out of mind, from time before mind, produce a dense, irreducibly complex intra-cellular creation to enrich us with precise packets of energy from the sun, from the early universe, from the world.

The upcoming pages will describe how we guide into our own life cosmic energy which is also the flow of entropic expansion that is also moving galaxies. We are all eating the Big Bang. We dine at the slowly rotating barbeque of the cosmos. Meat eaters or vegetarians, we are all creation-a-tarians.

Chapter Fifty-Four

Improbably Elegant, and Utilitarian

Now we need to dive inside our own cells and journey to the sites where carbon-chemistry food becomes our personal energy. Mitochondria, intracellular places that are half us, and half guests within us, are those sites.

Mitochondria, like the chloroplasts in green cells which contain chlorophyll, are thought to be endosymbionts, (Chapter Fifty-One) tiny lives once independent, but now gaining from and giving to our cells. Mitochondria and chloroplasts both contain their own DNA, distinct from cellular DNA, and exist inside their own semi-permeable membranes. They seem to be self-contained little beings themselves. Mitochondria breathe for us; we have the energy we do because we teamed up with them so long ago. They were once free-living little beings who now reside in our cells. Different human cell types, like liver, muscle, or brain, contain more mitochondria because they are more active, while other cells, like those that form tendon or bone, may have fewer mitochondria. With hundreds to thousands of mitochondria in each of our trillions of cells, we must have quadrillions of mitochondria at work every moment. Our life force pours into us from mitochondrial hordes where we harness energy.

Our energy is released inside the mitochondrial membrane on what we have referred to as a radiant scaffold. There, a ladder of enzymes dismembers sugar until only the carbon dioxide gas is left and is breathed out. In that breakdown of sugar, energetic electrons that had been holding sugar together are snapped onto specialized molecules that we possess in our mitochondria for the very purpose of carrying electrons. These electrons are passed from one specialized molecule to another, in an exact sequence, as they appear on the mitochondrial membrane, fifteen hot potato transfers of energetic charge and life force, a scaled and refined descent. On this radiant scaffold, enzymes carefully tease apart molecules, and release the ancient and original energy of the universe that had been temporarily stored in inter-atomic bonds that used electron arrangements.

Inside the electron carrier molecules, which are built mostly out of carbon, oxygen and hydrogen atoms, there are also rare atoms of metal. As we saw in photosynthesis (Chapter Fifty)—or as is commonly assumed when electricity flows through a copper wire—metal atoms that have a charge (metal atom "ions") are excellent at holding or transmitting electromagnetic charge. Life, our life, our bodies, have absorbed and skillfully sequestered the ions of several metals, so that, though they are only a trace percentage of us, metal ions are situated at the proper place on the energy-releasing, electron-transport scaffold, inside quadrillions of mitochondria.

There is no real division between "life" and "not life." We are the "light-metal" musicians. We are dead matter that has come to life over billions of years through information, organization, interaction, and evolution.

This is another moment where life can be understood as strategic location of atoms at profligate numerical dimensions. We are vast in number and minute in scale. We not only harness the energy of the universe, but we do so by sequestering and manipulating quadrillions of tiny atomic chips of metal ions that were spit out of exploding supernovae billions of years ago.

We know that DNA's information sequences tell amino acids where to line up, to assemble the proteins that form the radiant scaffold, which mitochondria use for electron transportation. It is something else that the proteins curl around and hold ionized atoms of iron or copper like sizzling wands that actually hold the hot potato electrons as they move along. These active catch-and-pass-on sites of electrons in the electron transport chain, which harnesses our energy from our food, are among the most improbable, elegant and utilitarian inventions of the universe. The radiant scaffold of electron-transporting, energy-extracting, protein enzymes in our quadrillions of mitochondria are unpredictable, unimaginable, and perfectly useful.

More than sixty protein catalysts cooperate, acting smoothly in series, in each of the quadrillions of our mitochondria! In many small steps the stored energy of the former food molecule is released and rebuilt into *new* chemical bonds, one after another. The electron transport chain directs, dampens and controls electron motion in service of energy release. As we saw in Chapter Forty-Eight, living electron transfer is always stepwise and constrained. Electrons are moved, chemical bonds form, and then the electrons move again into yet newer chemical bonds. Proteins form the scaffold, and metallic ions gripped inside the proteins are the baseball mitts, or the potholders, where the charged electrons which move along the transport chain actually touch down, hold, and then release to move further on down the chain.

Let us pause in wonder not only at the brilliance of life, but at the brilliance of the teams of scientists who have revealed these details.

In quadrillions of locations inside us, electrons are shepherded, carrying precious energy of the universe in infinitesimally small aliquots in innumerably vast herds at lightning speeds. *Electrons are flowing through us.* Our DNA-coded proteins are directing them like theater ushers. The matter-energy changelings, electrons, are in tightly orchestrated motion, giving us warmth, flexibility, awareness, life. Meet your maker.

Chapter Fifty-Five

Multidimensional Impossibility

A n example of the refinement of electron transport in service of respiration inside our mitochondria is the protein cytochrome oxidase. We can safely use oxygen only because of this protein's evolutionary emergence. Here we will find an even subtler expression of our own skill at befriending the Mercuries of the universe, the fourteen billion year old electrons. Respiration carries us into deep biocosmic creations. Cytochrome oxidase is a protein enzyme that is part of the radiant scaffold, and that is designed to handle the "oxygen problem."

First we need to pause and consider oxygen's promise and problems. It is an old element originally manufactured from nuclear fusion in now deceased suns. The element, oxygen, is made into more stable *molecular* oxygen that contains two atoms of oxygen mingling their electrons. Most molecular oxygen around us today comes from photosynthesis.

Molecular oxygen is a very active electron catcher. This accounts for the famous oxidative reaction that browns the apple which "oxidizes," or slowly burns, at room temperature. Oxygen is actually dangerous to life, because it is more than an electron catcher; it is an electron stealer. Scientists believe that as oxygen levels rose on Earth long ago, many simple life forms were killed off. A new breed of survivor was necessary to live among the higher concentrations of oxygen, the reactive. Many bacteria die when exposed to oxygen and can live only in deep soil or oceans beyond its reach. Oxidation of our own intracellular molecules probably is a major contributor to our aging and death. This is the oxidative stress theory of aging. Oxygen destroys many molecules and tissues by stripping off electrons like a thief, hence the famous antioxidants, that many people today take as dietary supplements. This also means that our energy extraction from food increases when oxygen is used as the final electron catcher. Oxygen is our deepest hole, the stickiest place to anchor at last the electrons flowing down our mitochondrial protein chains. Oxygen is the deep resting place where, if you put your electrons there, you have gotten the most energy that can be gotten out of them. But it is a dangerous friend, that can kill the same cells it helps.

To use oxygen, to harness its strong pull on electrons and the large amount of energy that this downward pull on electrons generates, life also had to evolve a method to dampen oxygen, to prevent it from randomly oxidizing (stealing electrons from) many other molecules in the cell. An oxygen molecule that has already stolen a single electron gains even more power to strip off still more electrons. As soon as you use it, oxygen becomes incendiary, an apt image, since burning is in fact rapid and wild oxidation, an example of the very thing our cells cannot tolerate.

Enter cytochrome oxidase, our internal mediator, the courteous bouncer among complex proteins. Cytochrome oxidase protein is an example of the long tapering exactitude out of which our own lives descend.

This large molecule, the equivalent of over two hundred thousand hydrogen atoms (greater than 200,000 "Daltons") in size, is a complex protein, built up from thirteen separate chains of amino acids wrapping around each other! To achieve its dual function of allowing oxygen to pull electrons deep and hard, while preventing wild oxidation of the wrong molecules, cytochrome oxidase needs a key: metallo-catalysis.

Proteins are made out of carbon, hydrogen, oxygen, and nitrogen— nothing special—nothing that can grip, use, and tame oxygen. Proteins are themselves vulnerable to destructive oxidation. Amazingly, cytochrome oxidase contains a metal vise. On one side an iron atom, and on the other side a copper atom, actively and strongly clench oxygen until it has captured from the food molecules all the electrons it can hold (four for each O_2 molecule), after which it can be safely let go, bloated and tame, as our most innocent friend, oxygen full of electrons attaching it to hydrogen, in other words, water. (Two oxygen atoms, each able to hold two electrons, form the O_2 molecule, that can therefore capture four hydrogen protons, to create two H_2O.) Of all the steps in energy production from the oxidation of our food, this step, with its dramatic thudding of electrons by four into the catcher's mitt, releases the most energy.

Let's keep in mind the fact that we contain quadrillions of mitochondria, each one of which contains cytochrome oxidase. As I write this, and as you read these words, quadrillions of molecules that are hundreds of thousands of hydrogen atomic units in size, are simultaneously utilizing and taming oxygen to release most of the energy by which we are writing or reading. We have to ask ourselves both how this numerical gigantism inside us evolved, and how "unliving" things, like iron and copper, became enlisted to serve us with such precision. The existence of such a large, complicated, and exacting molecule in us (and in slightly varying forms in all oxygen-breathing lives) numbering quadrillions per individual in uncountable trillions of lives on Earth, draws upon the regularities of the universe in the zones we call physics and chemistry, which create the atoms and their inter-atomic molecular bonds, and then layers over that the realm of biological wisdom. The amino acids that form the "letters" inside the protein "sentence," were themselves coded by the alphabet of chemicals in DNA. But to demonstrate what has been repeatedly referred to as "layered complexity," the DNA that codes the protein, cytochrome oxidase, has even one more added, unexpected dimension.

Multidimensionality is a deep recurrent insight into life. Sometimes we refer multidimensionality to well recognized domains, like physics, chemistry, or biology. But the more carefully we look, the more dimensions appear, *layered even within domains*, like the song that describes flies that bite the backs of the necks of other flies, and have flies that bite their necks also, ad infinitum. Now, within biology, we find a new layer of intricate causality in the DNA information that codes cytochrome oxidase.

Remembering that our mitochondria evolved from endosymbiosis, we can understand, while being staggered by the discovery that cytochrome oxidase amino acid sequences are coded for both by DNA within the *cell*, and also by DNA in the *mitochondria*. How did these two separate informatic memories not only combine, but cooperate? Cytochrome oxidase is not built from a cellular template, but from two different, separately evolved DNA templates that do not directly communicate. The actual active site of oxygen-gripping in cytochrome oxidase derives from mitochondrial DNA, while protein sequences that are coded by cellular DNA seem to play a more supporting role. All of this complex memory, and then precision sequencing of protein, must be accurately replicated quadrillions of times or more (since cells die and need to be replaced).

Aren't there a lot of errors when a sentence is copied quadrillions of times, particularly if that sentence needs to be simultaneously copied from two different books? Remembering that the protein sequences must somehow fold around, and grip, both an iron and a copper ion, which are not put in place by DNA codes at all (only RNA and protein are coded for by DNA, never trace metals), and that the proper position of the metallo-catalytic site requires transportation of the trace metals to the right place at the right time, and that the function of the whole is to safeguard against the dicey and dangerous oxygen stripping the wrong electrons from all the thousands of other molecules packed in the cell, so that there is no room for error, we get a glimpse of the layered complexity at the biological level alone. Yet this is a glimpse only. There is more for us to consider! The multidimensionality of life keeps luring us deeper into layers of problem solving and chemical inventiveness that accreted over billions of years. As soon as we think that we have discovered something deep, a deeper layer appears.

We have a very severe chicken-and-egg problem.

How did cytochrome oxidase (or its DNA sequences in two separate genomes) ever evolve? If its role is so fraught and exacting, then as it was slowly evolving, wasn't it initially inexact, error-prone, partial in structure and function? And wouldn't that inexactness in such a vital role lead to the unregulated oxidations that cytochrome oxidase is there to prevent? Before such a weirdly eccentric and function-fit entity as a bi-metallic vise could be first used, re-used, and perfected (over a million years? a billion years?) how could it operate in its half-baked, half-evolved form, well enough to even be re-selected by natural selection? Oxidative respiration, with its attendant danger of wild and random oxidation, seems difficult to create without creating it all at once, complete and successful, as it appears today, but this would seem logically impossible. Why wasn't the partially evolved cytochrome oxidase of the past oxidized and destroyed by the oxygen it tried to grip in its partially-evolved but still not well formed bi-metallic vise?

Chapter Fifty-Six

The Bridge

We are looking at the transformation of energy flowing through life, and how we and other oxygen breathers use the products of photosynthesis, sugar and oxygen, to break down sugar and extract energy from the electrons that hold together sugar's carbon bonds. Then we have selected one molecule used in respiration (among tens of thousands in the cell) to exemplify "layered and woven complexity." We have come up to an obvious problem.

We have been discussing a protein that holds two metal ions of two types, with no room for sloppiness or error. The amino acid sequence of the protein is coded for by DNA, but the critical iron and copper of the vise are not derivable from DNA code. Even presuming that trace metals are adequately represented in food (though no one eats copper wires for lunch, so that assumption is not certain), we still need to understand how components that are not told by DNA where to go, end up at the right place. Remembering that "law" means limitation and designation, what ordinance do iron and copper obey when they form their oxygen-holding vise? Who or what told these metal slivers where to go to protect cytochrome oxidase and the rest of the cell from oxidation?

From the start, life is always life: fluid and dynamic. Due to our conceptual limitations, due to our linguistic limitations, we frequently refer to life through static and mechanistic concepts. I have referred to energy flowing through life's "architecture" and "canals." Biologists frequently refer to elaborate cellular components as "machines." But these mechanical metaphors are misleading.

Life is always motile and changeable. Life's parts are vital when they are embedded in their living context. Vitality means adaptive *responsiveness*. It is not just whole organisms—animals, plants, people—who are ceaselessly in states of adaptive change, but even the molecules of life

236

within cells, pre-eminently proteins. Inside a cell, proteins are not sequences, globs, or machines. They are life living life. The shapes a protein takes are as important as the shapes of an elephant's trunk or a primate's hand. We need to think about protein plasticity, known as conformational allosterics. This will then guide us into understanding how cytochrome oxidase might have built its metal catcher's mitt.

Proteins have three-dimensional shapes, called conformation. Although proteins are coded by DNA (via RNA) into a linear sequence of sub units (the amino acids), once the string has formed, it folds, or snaps together into a curled, self-stabilizing shape. These secondary shapes are temporarily stabilized by soft types of chemical bonds that are more easily made, more easily broken, and weaker than the covalent bonds that structure the stabler parts of our inner world. One could use as an image here: a long, multiply-coiled strand of rope left in the garden; or a big, old, tangled grape vine. This coiled, three-dimensional shape, superimposed on the linear amino acid sequence, can also change. The changes in shape make proteins variously active, inactive, relating to their brethren and kin, or dormant. Protein conformation is "allosteric," multiply shaped. Proteins are not like sentences in books. They are lively snakes, grapevines in the wind, shape-shifting and plastic. In response to changes in cellular environment, such as alterations in the shape of neighboring proteins, or the presence or absence within the cell of various atomic elements in various concentrations, protein shape and activity changes. Proteins turn "on" or "off," based upon changes in their three dimensional conformation.

We are both rigid and fluid. Protein sequence is fixed, but protein shape is dynamic, responsive and communicative. Werner Loewenstein (see Chapter Three) describes the richness of shapes that proteins can assume as "an inexhaustible repository for inscribing information." Proteins are listening to, interacting with, and exercising some control over everything else in the cell. Proteins in a cell form a dense hive of mutual listeners and regulators. Proteins are citizens of the intracellular protein society, and they are alert and responsive to whatever is blowing in the cellular wind.

Conformational change in protein seems to be the rule, even the essence, of life. To understand this we need to readjust our sense of "molecule" to mean not a thing or an object, but a process, wave, oscillation, moving back and forth between three-dimensional states. Life is motion, on any scale. Proteins fold, flex and dance as an expression of life at the molecular level. Fish and butterflies share their style.

Protein conformation is so essentially "life," and so subtly nuanced, that it actually involves even deeper details. There is a group of cellular chemicals that have been dubbed "molecular chaperones." They are, themselves, proteins, usually small and simple ones. They guide and help bigger proteins to fold properly, like tugboats bringing oil tankers into a harbor. As we and our bodies face "stress"—events that knock us off kilter—our proteins, like cytochrome oxidase, can become shaken up, and lose their proper shape. One of our most important stress responses, that helps us preserve cell function and therefore our very life, involves the rapid increase in stress response proteins, molecular chaperones, who rush around the cell monitoring and adjusting the conformation of their bigger kin.

It is not enough that we contain physics information in our electro-magnetic atomic bonds, and biology information stored through evolution in our DNA. To survive the stress that life entails, we also need to repair, restore and buttress ourselves with little chaperone proteins who carry brilliant rectifying information from one hidden part of DNA to the dam-aged proteins that derived originally from another part of the DNA. *We are the inheritors of billions of years of adaptation and resilience.*

We make molecules to guide and heal molecules: we could say, "doctor molecules." These chaperones, tugboats and doctors in our cells enable the delicacy of protein conformation to endure in a world of chang-ing temperatures and many hard knocks. (Let's remember that protein chaperones, like all proteins, are coded by DNA, so, through them, DNA regulates its own products, the bigger proteins. This is an example of DNA self-interacting and self-modulating in woven complexity). When proteins are too damaged to be repaired by their chaperones, they are broken apart and recycled. Damaged proteins can clog up cell functions. Impaired re-cycling of damaged proteins may lead to diseases of cellular degeneration, like ALS (amyotrophic lateral sclerosis).

Now we have a potential answer to how copper and iron get placed properly inside the catcher's mitt of cytochrome oxidase.

Guided by conformational allosterics, the relatively large cytochrome oxidase protein, full of its inscribed information and skill, gains the subtle-ty and specificity to interact precisely with tiny atoms, like iron and copper.

Proteins contain the wisdom to guide the grains of the inorganic universe into the roles they play in the living world.

It is through this *descending cascade* that life can *finger and guide the universe*: DNA sequences code the RNA that codes the amino acid sequences, which form the proteins, which assume secondary conformational shapes, which can change with circumstances and are therefore fluid enough to feel the atoms one by one. This cascade bridges the span between complex life and naked atoms. Proteins like cytochrome oxidase can touch, feel, pinch and pick up things like iron and copper ions. The bimetallic catchers mitt in cytochrome oxidase, which handles the oxygen problem, is picked up and put in place by protein sequences that fold around and hold metal ions as if the proteins were domesticated boa constrictors. Remember that cytochrome oxidase has an unusually large number of protein chains within itself (Chapter Fifty-Five).

This critical role of protein conformational allosterics is by no means unique to cytochrome oxidase. The exquisite selectivity available through protein conformational changes is part of the constitution of life on Earth. Each protein is in itself inscribed with some living wisdom. We probably contain about thirty thousand shape-shifting wise proteins.

Take as another example calmodulin, the protein that modulates calcium function. It is worthwhile for us to spend some time to deeply understand protein conformational allosterics, because this phenomenon is less a feature of life, and closer to its core: the bridge between the atoms of the universe and the life in cells. Life is fingers and touches: delicacy by quadrillions. To understand how we breathe, we need to fully appreciate life's complexity, inventiveness, multidimensionality, numerical gigantism, ancientness, and wriggly sensitivity.

Proteins can change shape in so sensitive a way, that they can pick up individual ionized atoms and hold them in the flux of life, a skill they practiced and learned during four billion years.

Chapter Fifty-Seven

Lightning-Fast Listening

I n this Section IV, we are adding to our sense of wonder about life by considering how each one of us is nourished by the energy of the universe flowing through our aligned atoms by electron transportation. For us, primate mammals on Earth, we are oxygen breathers and eaters of sugar and its derivatives served to us by our hosts, plants. Oxygen based respiration, which we generally take for granted, is a supreme insight that took billions of years to craft. Even to describe respiration in the simplest way leads us down, layer upon layer, into the ingenious complexity of our existence.

As we break apart our food molecules and release electrons from the chemical bonds, we drive the electrons down towards our waiting enemy and friend, oxygen, and in accomplishing this incredible biochemical feat, we need cytochrome oxidase, which grips metal ions, and whose own existence (within mitochondria, in vast numbers) requires the uncanny ability of proteins to pick up and place individual metal ions within their atomic fingers.

Every cell and all of life is a fusion of information and skill. This is what we are about to witness in still more protein behavior. Proteins are sensitive and rapid at orders of magnitude beyond imagination. We will use as an example a well-studied and common protein, calmodulin, which helps to regulate calcium ions in cells, and our goal will be to realize more deeply the ancient and intricate invention we are.

While calcium has its famous structural role, forming bones, this visitor inside us, calcium, was formed in the furnaces of dead and exploded suns, as are all of the elements other than hydrogen and helium. Calcium is widely employed in our bodies to send signals, our favorite cellular messenger, for which it is ideal.

Calcium atoms readily exist in two forms. As an ionized atom, calcium stands free, containing a positive electromagnetic charge; or it can use its charge to bond to and participate in complex molecules. If calcium's charged ions flow across a cell membrane, the electromagnetic charge on both sides of that membrane changes along with the flow. The voltage difference inside and outside the membrane can in turn influence any molecule inside the cell that is sensitive to electromagnetic charge. Most are.

So flows of calcium ions across cell membranes can be utilized to change many functions of cells by changing the charge-level within the matrix of the cell. Similarly, any process, any other signal, which leads to the flow of calcium ions, which changes the function of a cell, can serve as a message. Signal-controlled release of calcium ions, to flow across a cell membrane, lies at the root of many of the processes by which our trillions of cells communicate with each other to form an entity, and by which one part of a cell communicates with other parts. The movement of signals in nerve cells, by which we voluntarily raise an arm, or think, is based on calcium ion flow. Muscle contraction is triggered by calcium ion flow. Often, biological signals are multiple and complex; calcium flow may form a critical step in releasing other signals. We are bigger, multi-cellular, and allegedly smarter than amoebas because we know how to release calcium ions in the right amount at the right time in the right place. Signals mediated by calcium flow connect us up to ourselves.

But just how do we release calcium ions? Released from where? And once they have been released, why doesn't that signal (triggered by the release) go on forever? How is the signal ended, or reversed? And then how do the storehouses of calcium get recharged for future signaling? And just how many calcium ions are we talking about? One? Tens? Thousands? A key player in answering all of these questions is calmodulin protein.

Estimates of the amount of one molecule throughout our body of course cannot be precise. Some of our cells may have as much calmodulin as one percent of all of the tens of thousands of proteins combined. Another estimate gives us ten million molecules of calmodulin per cell! Here we go again, because ten million per cell times one hundred trillion cells means an estimated sextillion calmodulin molecules in our personal possession, reminding us that we misperceive life if we limit ourselves to numbers and dimensions with which we are familiar and which make sense to us.

Life can make sense to us only when we can allow it to no longer fit our sense of "sense." We cohere out of a magnitude of manyness that is nevertheless integrated and conversant.

Calmodulin in its sextillions can use conformational allosterics to curl around and bear hug four calcium ions. When calcium concentrations rise, calmodulin can mop up hordes of these infiltrating atomic ions; and as calcium levels decline, calmodulin's boa constrictor conformation can relax, re-releasing a new supply of calcium. Calmodulin can modulate calcium flow and storage partly due to its agile conformational switches, but even more so due to its *sensitivity* to the presence or absence of its partner ion. Just how sensitive can a protein be? One estimate is that proteins on cell membranes (where their activity can be more easily measured) can open and close their shape in *a billionth of a second.* Calmodulin can hold and release calcium, and can also act as delivery boy, bringing the positively charged atom to where it is needed, at high speed.

This incomprehensibly *numerous, flexible, sensitive, lightning-fast, specific,* motile modulator of who we are acts in cooperation with many other features of our calcium signal regulation, such as pumps that move calcium into or out of cells.

Calmodulin reveals how our proteins shepherd herds of metal chips which in fact are shards of ancient, blown-up suns. As we saw with cytochrome oxidase, proteins can pick up invisibly minute chips of the universe with precision. When we eat with chopsticks we are imitating a universal principle of life.

The story of calmodulin parallels and supplements that of cytochrome oxidase, and shows how proteins are essentially fluids, and how life essentially grips, uses, and awakens isolated atomic ions, and incorporates them into the vast, responsive *communication network* we call life.

Notice that while calmodulin, like all protein, is coded by DNA, its conformational allosterics are also responsive to, we might almost say, under the control of, calcium ions. Calcium ions reshape calmodulin which regulates them. The biological, DNA-coded information in calmodulin is supplemented by environmental information that is constantly changing: that is calcium levels. Calmodulin is not only sensitive to calcium, it *learns* from and *responds* to calcium. Proteins listen and communicate. They respond to the ionic flocks that they regulate. They are good shepherds, cosmic border collies.

One definition of life with which we started is strategic placement of octillions of atoms (Chapter Twenty-Nine). Another aphorism to capture the wonder of life might be: life is a collection of proteins, which are coded for by the historical adaptations preserved in DNA sequences, and which transform sequential chemistry into skillful, precise shape-shifting in service of interaction, communication, and regulation.

Protein shapes may also be regulated by RNA independently from DNA, even though, in the past, RNA was believed to be a mere servant molecule. Cell biologists at Stanford have shown that varying RNAs can alter the rates of protein formations, which in turn alters their shape. Colluding and competing information sources, shifting rates and rhythms, and alternating bending motions all contribute to the attunement, adaptability and endurance of life.

In 2007 Neal Woodbury and a team of scientists reported in *Science* that even photosynthesis depends upon changes in protein *shapes*. Chlorophyll and the photosystems don't just stand in place like British soldiers to catch sunlight. They subtly *tilt and bend* to have an ideal angle, based on protein shape changes "on the timescale of a millionth of a millionth of a second." Protein conformational allosterics help green cells catch sunlight like shortstops. This point is becoming clearer and clearer: Life is three dimensional, in constant motion, and very fast.

Yet the *fluidity* of protein shapes seems to contradict the *precision* of protein action. If the shape is so critical, then how can it keep changing? Nor are all the mechanisms involved clearly explained by invoking DNA codes. What is the exact relationship between the DNA coded sequence of amino acids in proteins, and their snaky, multiple options of environmentally responsive shapes?

Although textbook biology generally states that the DNA code, which gets translated into the amino acid sequence, contains all the information that is necessary for the protein to fold into its mature and functional state, this dictum probably contains two errors. We have already seen how various protein shapes can precipitate from the same amino acid sequence. This shows that more than one protein shape is consistent with the same DNA information. More importantly, at each stage, as we have considered causality, we have seen *loops* and *networks* and *multiple causes*. Calmodulin is told what to do by its amino acid sequence, but also by the three-dimensional information contained in the calcium ion. Information is both central, flowing out from DNA, and environmental. Calmodulin tells calcium what to do; calcium tells calmodulin what to do. They are married. The little chip from the universe gets to boss around (a little) one of the daughters of life. Life has internal information in its DNA, and listens to environmental information, too. Even proteins keep abreast of the news.

Life always receives and responds to input. Cells and people, we listen to what is going on around us. Proteins change in response to conditions, and therefore conditions are information that proteins absorb. Life is not only coded from within but also informed by contact. Proteins can pick up and add (depending upon their shape and function), copper, calcium and other additional parts.

The essence of life is to modulate centrally coded information with periph-erally received information. Life is awareness and response. Life is always touching and listening. A one way flow of information, from a central storehouse outward, can't account for the conduct of calmodulin, which grasps and lets go of calcium ions in service of communication. Life listens and speaks *by touching, feeling, holding, releasing.* Changes in protein shape "R" us. We are all—gorillas, Super Bowl fans, sharks, and worms—very sensitive.

Chapter Fifty-Eight

Eluding Obstacles and Forging Connections

In order to understand the wonder of how we harvest our energy from the universe, we have been studying layers. Respiration is the way we get energy from food, and it depends on properties of oxygen chemistry. Oxygen chemistry provides the endpoint for the electron transport chain. The electron transport chain uses cytochrome oxidase molecules, which use metals in their catalytic process, and to hold these metals, proper protein conformation is essential. Calmodulin has helped us understand how protein shape-shifting combines central information sources with environmental responsiveness.

To fully realize life, we need to think not only about information but about *interactions*. DNA sequence prefigures any adaptation, but receptivity to the environment actualizes it.

Who could have dreamed this up? Why not use two copper ions in cytochrome oxidase? Why not twelve instead of thirteen amino acid chains (Chapter Fifty-Five)? Does the linear sequence of amino acids always take the same snaky shapes, the same three or ten shapes? How many allosteric conformations does cytochrome oxidase move through as it performs its tricky task? Ancient solar refuse, like copper and iron, sits so tightly and precisely within us. (Even the rare element cobalt is incorporated into our Vitamin B12's co-enzymes). How did living systems ever figure out that metallo-catalysis was going to be a useful tool? For something to be naturally selected, it has to already exist. How or why was a copper ion ever placed into the folds of the thirteen amino acid chains of cytochrome oxidase?

We have to stretch beyond domain-specific science to understand the existence of cytochrome oxidase as a function of multiple zones of law: two separate ancient DNA codes (mitochondrial and cellular), careful replication and translation of DNA codes into amino acid sequences of protein;

fluid, three-dimensional protein conformation, the proximity of organically sparse iron and copper, all in proper location inside the oxygen-using, electron transporting mitochondria. To some extent it is just mind blowing.

That is why we need and feel the sense of wonder. We cannot reduce life to comfortable concepts and soothing beliefs. There is something cosmic and charismatic about it that both startles and destabilizes us as we observe its billions of years of eluding obstacles and forging connections. Life has *preserved and transmitted solutions to problems,* ongoing sets of working answers, responses to existence, that bridge domains extending from the zone of a single atomic ion to an entire planet.

The inventiveness and precision of cytochrome oxidase that we have considered so far do not yet exhaust our journey through its woven, multiscientific complexities! More interactions of energy, matter, information, history, memory, and communication are necessary for our self-understanding, using cytochrome oxidase as our dramatic case-in-point.

Are you getting tired? Is this layered complexity and detail overwhelming? Pause, but don't stop. Remembering that Walt Whitman only glanced up at the stars from time to time (Chapter Two), take a break, but come back to look again. The deeper we probe, the more life expands around us with wonder. We have received from our culture the brand-new gift of being able to understand ourselves with cosmic and subatomic eyes.

In order to understand how we extract energy from food by transporting electrons from atomic bonds that we break open, we have had to think about proteins that translate the information of DNA into skillful *action*. Don't forget, we are talking about how *we move little bits of the universe around*. We are talking about *quadrillions of events and billionths of a second*. And we are talking about a sense of wonder at the limits of our knowledge.

Now we have to reconsider the fundamental question of causality. Standing on a cliff, we have to peer over the edge of law and order to observe its friend, randomness. The way we extract energy from food, the way we use proteins and move electrons, plunges down into the deepest layers that causality offers. Life has capitalized upon the ethereal zones of quantum probabilities. We contain skills that we could never have dreamed up. Life is strictly unbelievable. We have been born from billions-years-old gestation in a womb that held every skill that is available to the universe's inventive combinatorialism.

Chapter Fifty-Nine

Tap Dancing and Dovetailing

Physics, chemistry, and biology refer to realms of regularities, reliable probabilities at different dimensions involving different phenomena. They are all players on the ball field of energy use in our cells. There is yet another realm, a product of motion and configuration.

Chaotic lawfulness, statistical probability, also enters in. The elements and molecules of which we consist, which we make and break, and the energy whose flow we are, cannot always be grabbed and held like a bit of copper or iron. In the essential fluidity of life, we also spring up from tap-dancing collisions. Not all of who we are derives from the etiquette of formal dance. We are not entirely products of DNA libraries with their preservation and erudition. There is a bit of dance-spree in us as well. Inside the deep order of life are things just bumping into each other. Alongside the unimaginable precision of cytochrome oxidase, the electron transportation chain of energy extraction also exploits probability, randomness, goodness of fit. Here is how:

Proteins like cytochrome oxidase regulate electron flow partly by *location*. For example, the many electron transfer steps releasing the energy of our food all occur side by side inside the mitochondria. But proximity is not destiny. We do not always get along well with our neighbors, or even our relatives. So electron transfers in service of energy release also use another process for regulation, and that process is the fit of detailed chemical *shape*.

Life's activity is three-dimensional. Shapes fit shapes. Electron transfers (and other life processes) follow hand-in-glove placements. Just like with boys and girls, if the chemistry isn't right, there's no consummation. Electron energy transfers occur not only due to position, but also due to constrained random diffusion that is at last curtailed and guided into place by shapes that fit at their nuanced chemical edges. Molecules interact effectively when their active sites, their subtle shapes, dovetail to high levels of precision. Not all molecules in the cell are anchored tightly into place, and as the loosey-goosey ones dance around, they meet and mate, and can do chemistry, only with those other molecules that fit them with atomic perfection. It's the details that make love, even among chemicals. *Surface alignments* permit electron transfer among protein catalysts at *one place only*, and preclude all other places. While we tend to think of chemicals as formulas or as "stuff," when we descend into atomic realms, every atom is a space-occupying configuration that can either fit into or collide cumbersomely against another. All chemical activities of our cells, including electron transfers to move energy, fit together active sites of chemicals like the space shuttle docking alongside an orbiting satellite.

A remarkable example of the high specificity of surface alignment is quantum tunneling, which uses probability and motion to create stability and order in the cells of our bodies. Quantum tunneling reveals that we are partly products of random oscillation coupled to specificity of active sites. (Quantum tunneling was part of Philip Anderson's original discussion of more becoming different.)

To understand quantum tunneling as an example of electron activity, we have to remind ourselves about the attributes of electrons. In their own subatomic domain, electrons are located in a "smeared out" way (Chapters Seven and Forty-Eight). They are not exactly in one place, but have the potential to be located in particular *zones* rather than exact spots. Somewhere within the protein's hollow, an electron can be held as a *probability of location*, rather than as a solid golf ball. The subtly located electron is more or less likely to be found in certain sequestered pockets of the complex architectural folds inside the protein. Vibratory probability allows the electron to be hidden within the protein, or exposed for interaction, on a likelihood basis, that is frequent enough to be utilitarian. Interactions that use the electron's energy may or may not occur, depending on when and how the little dancer appears in front of the protein curtain. Changing the protein shape can *change the probability of electron exposure,* to regulate the rate of an interaction which uses that electron.

The tiny, vibratory oscillations of electrons, deep inside the snakily folded, three dimensional conformation of protein shape, is yet one more determinant of, and control over, whether and when an electron-driven chemical bonding event, will or won't be catalyzed by an enzyme. This quantum tunneling is the fanciest example of how the probability of shapes, fits, and surface alignments can determine chemical interactions in cells, as much as can actual structure and location.

It is such subtlety, elegance, and *multi-domain science* (quantum physics used by combinatorial chemistry for cell biology bioenergetics) that may partly account for what appears to be unnecessary and cumbersome aspects of biological molecules. Maybe that thirteen sub-unit protein chain, cytochrome oxidase, has active inner compartments that do indeed require so many entwined sub-units. Given its fielding skills, cytochrome oxidase's multiple folds may be part of its skillful functions, some of which may entail electron juggling on the quantum scale.

Like a soccer ball, always sort of in front of the player, but moving back and forth from foot to foot, and never precisely in frontal line, the electrons of life follow the laws of quantum physics and the laws of protein conformation, while the big chemicals transfer them with professionally honed kicking skills. Life, the precisely detailed, life, the citizen in the dictatorial realms of electromagnetism or carbon chemistry, is also a sportsman. Not only structure but also activity guides electron-flow. And electron flow is our most intimate essence.

The radiant scaffold of proteins that transfer electrons on down to oxygen may well hold, hide, or release the electrons by processes similar to quantum tunneling.

When we discuss detailed and apparently arcane science, like "specificity of molecular surface alignment," or "quantum tunneling," it is important to remember that we are a *community* of such processes, in the way that New York is a nest of traffic-conducting streets. These processes are how our life-giving energy flows within us. These terms are not topics of science merely, but discoveries and descriptions of *how we have emerged out of matter, energy, and information.* Could it have been any less arduous and inventive for mere elements to have cohered into people?

In summary, electrons in our energy extracting cells travel paths guided by proximity, paths guided by molecules, paths guided by metals, paths that are based on fit or flow, paths guided by pockets and tunnels. Electrons in our energy extraction processes also sometimes utilize probabilistic statistical mechanics. All of these varieties of paths have evolved to move electrons to make and break chemical bonds to transfer cosmic energy into cellular life. We are the product of multiple levels, variety, opportunism, skill, probability, communication, and coordination. We are inconceivably intricate and creative. Once again we are encountering a stage where if you feel you are beginning to understand the complexity of life, you don't.

In the oscillations and motions inside us the fundamental properties of matter have mutually engaged into self-regulating societies. Elements become molecules, and electrons stream into energy patterns. The final product steers the original energy, that got packaged into hydrogen, that bursts out from the sun, to flow in circuits of human life. On the inside, this is all regulated partly by the snowflake shapes of touching atoms.

Chapter Sixty

Herding the Ancient Ones

At last we can sanely and scientifically identify ourselves with the cosmos. As the great biology text, *Molecular Biology of the Cell*, by Alberts et.al. says, "Electron transfers provide most of the energy for living things." *Stepwise flow of energy through electron movement is the essence of life.* (When energy flows by alternative processes, such as resonance, it is nevertheless usually channelled back into electron motion.) This flow of electrons in our cells is part of the electron movement that began when electrons congealed in the early universe, even before atoms were born from the marriage of electrons and protons. Our own bodies' electrons predate everything we see in the compounded world. They predate the Earth, the sun, the stars in the night, the atoms of the world. Electrons are the natives. There are other equally old, or older, subatomic particles, like the quarks that aggregate to form the protons in the nuclei of atoms, but these other sub-atomic mites are hidden in the depths of things, and are not direct actors in our daily life, as electrons are.

The cycles of photosynthesis and respiration drive around the old electrons like herders guiding an antique breed across the Hebrides. Down our electron transport chain the ancient ones flow, and our eyes sparkle with the result.

To hold in mind the cosmic saga by which we receive our personal energy from our universe via photosynthesis and then respiration, let us trace schematically a mythic electron, "Ellie," through whose greatly shortened legend we will capture an overview.

This electron, "Ellie" congealed out of pure energy in the very early years of the small hot universe as it expanded and cooled. Over eons, Ellie rode hot sub-atomic plasma in explosive regions of the universe, and also found, from time to time, temporary homes in stars, where she married and lived with a proton as a hydrogen couple. To Ellie, stars are temporary

summer cottages. For a long time she lived in a big star whose gravitational force fused hydrogen protons into heavier atomic nuclei, and eventually Ellie lost her proton to fusion and became a widow. When the big
star had fused enough hydrogen to deplete its fuel, it exploded, and in the
supernova chaos that followed, Ellie bonded to a magnesium nucleus as
one of the twelve electrons married to the big, bigamist nucleus. Twenty-
four of them, twelve protons and twelve electrons, careened through space
(real magnesium also has neutrons) until they became bound by gravity
into our newly forming planet Earth, where in fact magnesium is common
in oceans and elsewhere. They rested a long time as newborn Earth ges-
tated, a billion years or so, until Earth began to cry with life, and Ellie and
her magnesium commune were absorbed by a "cyanobacteria," a simple
cell who was a pioneer photosynthesizer. Ellie was ushered by DNA and
clever proteins to a front row seat in a chlorophyll molecule in the photosystem of the green cell. (In another telling of the story, Ellie's twin sister
would have made her way to the photosynthetic hot spot via water. The
electrons that eventually participate in carbon fixation, as described in the
next paragraph, are from water rather than magnesium.).

On a radiant morning, sunlight streaked down and burst into the
chlorophyll, grabbing Ellie by the hand, and raising her up from her front
row seat and onto the stage. She was filled with the hip hop electromagnetic energy of light ("I want to celebrate my life…") and free of magnesium
(or water) she danced down, with the help of burly co-stars RBC, (Chapter
Forty-Seven) to stand and sizzle, a twirling ballerina, holding together
two carbon atoms who came "clubbing " from their carbon dioxide past
in the Earth atmosphere. She held carbon's big, clumsy hands, and by her
energy, carbon was fixated, and she was soon dancing in a chain of carbon
atoms, as sugar. The carbon dioxide's abandoned girlfriends went back out
into the air as molecular oxygen, O_2. Ellie and her new sugar family may
have cycled through the biosphere in various forms for billions of years (in
a staggering oversimplification that this story requires) until, as part of a
carbon-to-carbon bond in a sugar recently refurbished by a brussel sprout,
I ate Ellie, house and all.

With my lungs I grabbed oxygen from the air. My smart cells broke
open the sugar's carbon bonds, and Ellie, liberated from eons of carbon
domesticity, danced again with free energy twirling down a chain of protein stairs in a mitochondria in my biceps muscle, partnering in dance for
a fraction of a second with one of my quadrillions of cytochrome oxidases,

and dancing on until, thud, she tumbled down into oxygen. I was incandescent with the gift of energy Ellie bequeathed, and I wrote this sentence. Ellie, still attractive despite her fourteen billion years, put her negative charge against an equally mature proton, and I breathed the old couple out (as part of either carbon dioxide or water).

You and I have mingled our thoughts on this page due to Ellie's transport of energy. Ellie has been breathed in once again as carbon dioxide (one choice) by my maple tree, and, joining a photosynthetic team again in the green leaf, became a sugar dancer once more. Despite her venerable Big Bang pedigree, she was ignominiously eaten by an inchworm on the maple leaf, and provided energy for it to hump up once in the middle, and so Ellie journeys on, indestructible, electromagnetic, just as active as ever. She emails with old, old friends in other stars, even other galaxies, that dispersed in the early days from their common home in the ancestral, small hot universe.

Even the elements in the rocks must defer in seniority to the electrons in our cells. Our bodies are warmed by the sparks of the first cosmic lightning, and play marbles with cosmic grains. Traveling electrons, currently residing in our atoms, crossed the cosmic empti-tudes, and are the active citizens inside the equally complicated cosmos inside our skins. We have built, we are the product of, molecules that guide electrons to pour their speed into our bodies and minds. We have captured, cornered, guided and milked electrons, and it was electrons who gave us the power to do that.

Werner Loewenstein has pointed out in *The Touchstone of Life* (see also Chapters Three, Twelve, and Sixty) that all the chemical combinations and electron transfers of which we consist, occur at a very limited distance, which he has dubbed "the serviceable nook." Too far away, and the transfers and combinations just won't happen. Too close, and electrons from the two chemicals will repel each other, because in the electromagnetic constitution of the universe, two minuses (two electrons) repel each other. The serviceable nook, of which we are the complex expression, is a few units of a ten billionth of a meter! We are exquisitely refined and full of *gaps*. We emerge out of dictatorial laws and oscillations. We fit and we move in secret worlds, where refined sorting and random bouncing look alike. The spaces themselves move and bend, as do the structures that form them. What less would you expect of energy flow? We are matter like electrons, we are energy that electrons carry, and we are a trail map of tiny spaces where cosmic energy goes spelunking.

When we look at life we see that *a potential, latent within possibility,* has emerged. Out of innumerable possibilities, one has been actualized. Can we really say this is the only way that evolution could have adaptively responded to the problem of energy creation? Was the large, intricate, detailed cytochrome oxidase the only possible answer? Aren't we left standing alongside Lewis Thomas in our shared dazzlement of surprise? Of the many ways to make chemical bonds, break chemical bonds, release energy, use oxygen, is the sequence and three-dimensional, conformational folding of cytochrome oxidase somehow written into the universe, or emergent from the universe, or compounded and woven out of the realms of quantum physics, chemistry, and biology, inexorably, or capriciously?

If life has evolved elsewhere in the billions or trillions of planets that might sustain it across the billions of galaxies our telescopes see, would it use oxygen as the final electron receptor? Would the oxygen-holding catalyst for this energy-releasing reaction resemble cytochrome oxidase? Would quantum tunneling of electrons inside of folded protein chains be involved?

No wonder the origin of life, the meaning of life, the purpose of life evade rational and evidentiary conclusions. On the one hand, we living beings seem to actualize elegance, sophistication, and fine-tuning that is exquisitely specific and precise. Yet every detail about us, like the existence of our cytochrome oxidase, seems wildly arbitrary.

Out of nothingness is the emergent something constrained to flow down a limited set of potential paths, or is it pre-determined down one? Does the final common path of life require quirks of history to prune multitudes of possibilities down to singular actualities?

These questions are what my quadrillions of mitochondria empower me to wonder. Electrons dancing and proteins writhing in such large numbers with such exactitude are the preconditions of my questions. Are my thoughts of wonder really my own? Do electron jumps and protein allosterics limit my thoughts? Drive them? Necessitate them? Preclude their alternatives?

A great deal has to be in place before I can think, before I can wonder. Do all of the billions of years of compounded organized complexity *allow* me to feel wonder, or does it *compel* me to do so?

Our thoughts, which feel so "true" and "free", are all predicated on the motions of the particles of the universe break-dancing in ballets by the billions.

We may capitalize upon serendipity and contain path dependent derivatives of circumstance, but we are no accident. We are collections of long-nurtured solutions that have worked. It took a long time and a lot of editing to make every one of our molecules. As the offspring of such a long streak of inspiring successes, let's allow ourselves just a brief, momentary, "Yeaaaay!"

Chapter Sixty-One

Our Real Identity at the Nexus

Now, after so much detailed scientific thought, let's pause for a moment and review the big picture that we want to hold in mind, in order to experience scientifically compatible wonder.

The biocosmos moves electrons across chasms one ten-billionth of a meter wide, by the quadrillions, every second, inside my body. Our jeweled compaction is no less cosmic than are the galaxies with their big black belts. The amount of information and skill with which any one of our constituent trillions of cells is assembled outpaces our cognitive capacities with numbers, inventiveness, and precision. The more we learn about ourselves the more we recognize that our conscious identity is a shallow cap on our galactic physical assemblage. Our real identity is at the *nexus of the cosmos with the biocosmos.*

We have come to appreciate why it takes so long for complex life like ours to evolve. We typically walk around and worry about small things while inside of us the universe rolls its energy down lanes and avenues inhabited by great inventors, old libraries, and royal jewelers. The energy that animates us began with time and steers through our molecularly walled streams, awakening us into self-awareness.

No wonder we cannot capture the essence of life in our understanding. No wonder our life seems to evade definition, because its proportions are so large, its aggregates so numerous, its dimensions so small, its inventiveness so great, its lineage so long, its information so large and so compounded.

This great trick I do so well, so quickly, so often, of throwing electrons into oxygen traps and keeping for myself their dropped energy, what could more confidently establish me as woven from an inextractable overlap of laws from all domains? We all toss around octillions of atoms like

Frisbees. We feast on electron energy that we pluck from the tree of life, photosynthesis, and we breath out in respiration our slow and stately fire.

In this Section IV our journey of wonder started with photosynthesis. Green cells put together molecules whose component atomic bonds hold energy, and the same cells breathe out oxygen as waste. Humans breathe in that very same oxygen, and use its electron-stealing power to break apart the molecules that the green cells had just made.

As we break the molecules, they release energy, which we then re-package, into chains of molecules that move electrons and energy to the places that we need them to be. We build the living systems that are necessary for the systems to live.

Our electron-moving, energy-moving chains capitalize on the laws, the predictable regularities, of all aspects of the world, starting with the universe's origin, on down even to the dicey, probabilistic laws of quantum oscillations of electrons. At every stage of our own intracellular life, molecules are told what to do by the physics constitution of the cosmos, and by the biological law book that slowly was written down and preserved in DNA.

Our molecules obey an intricate tapestry of laws, and are manufactured in gigantic amounts, yet with detail that is precise to the size of a single atom, like a single copper ion. The large numbers and detail occur in a fluid context, for every molecule within us, and the collection of molecules that is us, change shape continuously. The sinuosity of living molecules dictates their varying and adaptive interrelationships, coordination, and capacities, even though they are all trying to touch down on and link up with both writhing and moving targets. The universe of molecules inside us must be kept from smashing into each other or from moving too far away from each other. Essential handshakes must occur despite the presence of a Yankee-Stadium- sized crowd of other molecules trying to shake hands on other deals. Tiny flakes of ancestral exploding stars, like copper, are given exact roles in exact allocations.

Electrons from the early universe move through this minute and infinite maze guided almost without error to flow through and charge life. The total picture is a mosaic, a weaving, a nimble interconnecting of atoms, energy, and laws from zones that we call quantum physics, physics, chemistry, biology, but the actual constraints and guidance—that is to say, the information—that comes to bear on any one atom as it participates in a molecule that participates in a cell, is as trans-calculable as is the number of atoms in the cell. Yet atoms are the bigger chunks, and it is at the level of subatomic electrons that we gain "that-which-moves," energy, from the expanding, exploding, entropic, billion-light-yeared universe.

The information in our thoughts is insignificant compared to the information in our whizzing becoming.

And we have not yet taken into account how the energy released by food or respiration actually gets to *where* we need it for the construction, for example, of another strand of DNA in the nucleus of the cell. How do we build ourselves? How does the energy released by the breakdown of food, by the electron transport chain in mitochondria, get to the construction sites where atoms are built into the smart and swarming molecules, like cytochrome oxidase, whose aggregate interactions we call life? By this point we know that the creation of life, molecule by molecule, will entail a tapestry of every realm of law, in unthinkably large numbers, with nano-precision. But there is more to stir our wonder.

We need to open up an entirely new Section to understand how we *construct* ourselves out of the matter, energy, and laws of the cosmos. So far, we have analyzed. Now comes assembly. We are complicated, but how do we become both flexible yet *unified*? How do we invent ourselves while we direct the inventor?

SECTION V

Microcosmos: The Creation Within Creation

Chapter Sixty-Two

The Swirl of Becoming

Wonder rises up when we become identified with our vast and infinitesimal context, and experience ourselves as *expressions* of the same processes we observe around us. We explain something about our own nature from the realm of the billion-light-yeared quasars, and from the realm of billionth-of-a-centimeter intermolecular gaps. We experience ourselves in continuity with the past and with the widely networked moment. And we recognize connections everywhere, because even when physical events cannot touch they share origins and regularities, which connect us up within ourselves and between ourselves and anything else.

When we open out in wonder, our experience becomes *embedded* and *participatory*. Everywhere we look, something is telling us about ourselves. We are being carried forward to wholistic apperceptions, the open-mindedness and the fulfillment of wonder.

While thinking about the wonder of life in Sections III and IV, we have defined it as strategic placement of atoms and as channeled energy flow. We have seen that our constituent atoms are ancient, many of them almost as old as the known universe. Our energy is even slightly older than our oldest atoms. While every one of us is therefore cosmic, we are also brand new compounds, absolutely unique combinations and permutations of ancient components. We are made not only of matter and energy and their fuzzy in-between, but we are also somewhat regular, predictable and lawful. We contain, or express, many types of laws and limits. When our atomic matter, our cosmic energy flow, and our lawful regularities are all subjected to very large amount of combinations and permutations, over a very long time, we emerge unique, singular specificities out of that "woven complexity." Our woven complexity is also "layered," sweeping up and utilizing laws from the realms that humans call "quantum laws," "physics," "chemistry" and "biology." Each of our component parts—atoms, energy, and laws—is so numerous and so compounded that we all represent "numerical gigantism" or "cosmic dimensions."

Even though we are each so primeval and manifold, we are also detailed, and so refined that we can skillfully poke around in the sub-atomic realms. Sub-atomic components, like electrons, organize our structural bonds, atom to atom. Electrons also form our energy flow, which is constrained and guided along the scaffold of eerily competent macromolecules. Our large molecules exist in vast galactic arrays, and yet they finger and place electrons with a jeweler's skill. We not only consist of flows of energy from the origin of time that have been channeled through hydrogen atoms and the sun, but we also operate in regions that are billionths of a meter small, and yet we do so with a minute error rate.

We download the cosmos into the rivulets of our microcosmos.

Inside us, everything is alive and motile, the "structures" themselves are rapidly built, destroyed, and rebuilt, and are always responsive, bendable, and interactive. Just as we communicate with the cosmos, we communicate with our own trillions of cells and octillions of atoms in our interconnected and creative network of bodily life. In spite of all that creation, we also obey the entropic principle, and eventually dissipate into the cosmos again, though often having first transmitted our biological or cultural information to others for them to capitalize upon. We are manifestations of both "compounded order," and of randomness, oscillation, and decay. We are part of the universe, and can't jump out of its circle nor escape our partialness. We are members, but we are temporary, so that at some level we are not only today's momentary interconnected order, but we are also yesterday's and tomorrow's far-flung swirl of becoming. We are brand new and recycled.

Intelligently refraining from grandiose, pompous, or self-appointed statements about what it all means, and avoiding concrete belief in silly fairy tales, we hold our lives and the universe from which they sprang in wonder. If we yearn for a moment of soothing ancestor worship, we turn towards the sun and green plants, our most immediate life-giving benefactors. Or we turn to the wonder that there is something, not nothing, the wonder of cosmic law, information, the wonder of existence, materials and patterns.

But wait! We have just finished examining, at the end of Section IV, how we harness energy from our food by transportation and transformation of electron energy. How does that energy ever get built back up into our human body and mind? The wonder of life not only includes the fact that we harness cosmic energy, but also the fact that we recreate ourselves moment by moment. We are builders, contractors who construct finely crafted humans out of antique particles and careful plans.

The fabulous catcher's mitt of cytochrome oxidase and its friends have released the energy of speedy electrons and delivered them over to oxygen. Oxygen full of electrons attracts hydrogen to form water, to restore to Earth cycles what plants withdrew. By these processes we have captured energy. But there are tens of thousands of large, complicated molecules in every cell in trillions of our cells. As molecules are constructed atom for atom by the trillions (or more), how is the energy that we obtained, from our skillful harnessing of electrons, applied to the many-fold and separate tasks at hand? A fantastical construction project remains unexplained. We have the energy, fine, but how do we apportion and apply it to the myriad creations of biosynthesis? And the communication networks to form an integrated whole, where do they come from? Atoms still need to be strategically placed, energy still needs to be apportioned, and the message of life itself, where and how to put in place the legions of octillions in their billionths-of-meter arrays, still needs to be spoken. Along with the energy of molecular construction, we also need energy for integrating the horde of parts into a functional whole.

It must be very clear now that we are almost entirely unconscious. For all of our understandable pride in our consciousness, which today has the capacity to look out at the galaxies through the orbiting Hubble telescope, or to perform X-ray crystallography on cytochrome oxidase and many other chemicals too small to see in a microscope, we cannot even vaguely make ourselves. Our consciousness is a last, late arrival. We are complexly alive with "woven irreducible complexity," "layered lawfulness," and "numerical gigantism" long before our first thought, long before we poke our heads out of someone else's birth canal. We are a little clever but infinitely innocent. We are products of time, history and laws, whose magnitude exceeds our thoughts, not to mention exceeding our manipulative and constructive mental capacities. Our thin stream of consciousness is a grand banner, but our life is sunk deep in a body within which its nature lies. Inside us are energy-moving messages in astronomical numbers and webs.

Very small messengers race here and there to bring us into being. They attend to rules, and even skillfully fence with subtle disarray, to weave life and consciousness out of the pinpoint energies and matter of the early universe.

Our life is orders of magnitude smarter than our minds. We experience our conscious selves as the chief executive of our bodies because we can voluntarily raise our arms, decide when to eat, or even think, and of course our volitional and cognitive capacities merit extensive praise. But our platform of living tissue is built by trillions of cells that uptake matter and mold it into life, that communicate to each other so swiftly, accurately, and knowledgeably that we, whole organisms, precipitate from coordination among the trillions. All of this is accomplished, not by our self–adoring consciousness, but outside the realm of our comprehension. What we call ourselves is only the snowcap on our own mountaintop. Our "self" is just the couplet that concludes an epic poem. Like water striders, we float on the surface tension of an oceanic history. We are a temporary message in time and space that summarizes findings from an encyclopedia, a dictionary, a library.

To feel the wonder of life, we need to watch as our own cells construct us out of molecules and energy, while playing with "BBs" of matter, and while rounding up the posse to protect us from the rough neighborhood of planet Earth. Our heroes in this Section V, on creation, will be the molecule, ATP, the process of proton pumps, and the splendor of our immunity.

Chapter Sixty-Three

The Coin of the Cosmos

To understand how we construct our cells out of atoms, energy, and law, we need to re-examine the phenomenon of quanta.

Quanta are patterns that we locate repeatedly across our observations of nature. The universe has its flows and continuities, its electromagnetic waves, and its planetary rivers; but the universe also seems to run on units. There are relatively fixed, regular discontinuities, lumps, or bundles, which make things work. German Nobel Prize winning physicist, Max Planck, initiated quantum physics with his discovery of a minute unit, or "quantum" of light, which he was able to enumerate mathematically with "Planck's constant." From Planck's discovery of an entity-like unit in what had previously seemed a fluid and seamless world, arose Einstein's particle-wave duality of light, the quantum mechanics of Heisenberg and Bohr, and modern physics. When we take this idea with us in our journey through life, we see the principle repeated. Can we say atoms, particularly hydrogen atoms, are quanta of matter, or that stars are quanta of the universe's matter? Or is it galaxies that are the quanta? Stars and galaxies are clumped discontinuities in space, but they are not exactly fixed or reliably quantifiable. So we need to look further.

Human beings spontaneously recreate the quantum principle by creating "pennies and dollars" to make the economy flow. Money reduces material exchange down to defined, exact, repeating units. Like light, money has dual properties, in this case as material and as symbol. We have seen that life has its unit, the cell.

In its deeper chemistry, life has a quantum-like use of energy, its "dollar bill," by which the economy of the cell can make easy energy exchanges. The energy of the universe, when it enters life, stops its flow, and moves in units, in quanta, that are contained in chemicals. This chemical energy unit is emblemized by everyone's favorite molecule, "ATP," the dollar bill of life (first mentioned in Chapter Three).

ATP is the symbol of life's unity, specificity, and inventiveness regarding energy use to build the molecules we are. (There are other, similar, molecules in its family, but ATP explains them all and takes center stage.) Every life form, every bug and every fashion super-model, uses ATP. Its ubiquity reveals that all life shares a common history and an identity. All life uses ATP; all life has a common origin, process, and bio-energetic. As a penny, or a dollar bill, or a Roman coin contains an imprint of a president or an emperor, ATP shows the face of life. It is imprinted with life's conservative continuities, and unpredictable, weird choices. Life on Earth as we know it is a particular something. It is woven, but it is not a mere blend. Like a visage on a coin, ATP expresses a *fundamental single government within all of life.* Every living cell on planet Earth—in the eye of a tiger, or a bacterium in my colon, or in my brain—is a citizen of the same government, and ATP is the coin of the realm.

When energy is used by a cell to make our molecules, and when energy is saved, put in the bank, for the next construction project in our blood or bones, that energy comes from, or goes back into ATP's chemical bonds (or the bonds of ATP's relatives). ATP is a central bank.

In uncountable, unconceptualizable legions throughout every cell in the biosphere, the energy of the wild, black-hole-drilled cosmos, is doled out of the pinching forceps of ATP.

The bull in the China shop, the exploding Hubble universe, places its energy into the billionth-centimeter "serviceable nook" with the eyedropper of ATP. Energy in living systems is doled out in a quantum that ATP controls. ATP is how the light-years become miniaturized into billionths of a centimeter, how the fiery sun moves my cool finger. ATP is the coin of the cosmos within the economy of life.

How did the unique invention, this pervasive tool, ever come about? How did it become so widespread? How does ATP, the maker-of-all, get made? What is ATP?

Chapter Sixty-Four

Extremely Rare Things Are Common

We are considering how our collection of cells is constructed, and how its ordered atoms and are maintained for a lifetime against the more probable entropic dispersal (to which we eventually succumb). The first idea is that life distributes packets of energy to the right place at the right time in the right amount, despite the fact that food energy is variable in amount and timing, and despite the fact that cosmic energy is irruptive and explosive. But before we can understand the quantum energy delivery system of ATP and its siblings, we need another background idea. We must return to the conceptual tool of large numbers. After taking some time to explore critical implications that large numbers have for life, we will apply those implications to the details of how ATP regulates the construction project that is us.

In Chapter Thirty-Seven, we saw how numerical gigantism in our bodies limits our intuitive self-understanding, since our conscious experiences do not produce salient encounters with numbers of the scale we need to think about life at its fundamentals. Now we need to consider an entirely different consequence of bio-numerical-gigantism.

The invention and natural selection of extremely improbable things, like cytochrome oxidase, (Chapter Fifty-Five) or like smart proteins such as "calmodulin,"(Chapter Fifty-Seven) or (as we will soon see) like the entire antibody system of immunity, which to me is the most counter-intuitive, extreme, and improbable surprise among life's dazzlements, all rest on the power of large numbers. The functional role of ATP within life also rests upon this numerical impact.

As events repeat, unlikely things become more likely due to the repetition. Repeat an event often enough, and its rare variations become probable. Something that would happen only one time in a billion will probably happen one thousand times in a trillion events, and will probably recur one thousand thousand times (or a million times) in quadrillion events. Extremely rare things—once in a billion—are common; we have already seen how our body requires quadrillion duplications of its original DNA (Chapter Forty). Every one of us, every organism, is a repository of the unusual becoming usual. *Numerical gigantism is one of the keys to life's inventiveness.* It is also one of the keys to understanding evolution, or to understanding our embodied selves.

By the term, "numerical gigantism" we are trying to capture elusively large dimensions. Can we then create a magnifying term, "compounded numerical gigantism?" We need to go to this extreme, in order to understand the most fundamental idea in molecular evolution. In one person, the original single cell divides approximately one quadrillion times, but there are billions of people. Even with a very low rate of replication errors, still, the base number of replications is so huge that every single one of the three billion nucleotide letters in human DNA will have been mutated numerous times over numerous years. The gigantic number of cell divisions guarantees that a very large number of subtle errors will also occur. The compounding of cell division in large populations over long time-periods gives evolution the opportunity to try out every possible variation of DNA repeatedly. Evolution is not about trial and error, but about many trials and many errors (therefore we could abandon the term "error" and say instead "non-causal creative variation"). Every living being, each one of us, is like a roulette wheel where the chancy bet will definitely sometimes win. The biosphere is a cauldron in which the unthinkably improbable will definitely bubble to the surface. In biology, in ourselves, the odds "one in a billion against you" is a guaranteed winner. This is why, eon after eon, despite smashing movements of the Earth's crust in plate tectonics, despite asteroid impacts and mass extinctions, you can bet on life.

Let's look at how this *certainty of the improbable* operates in our life. Out of our trillions of cells, many exist in dynamic tissues like blood or skin, in which there is rapid proliferation, death, and replacement. We each suffer billions of cell deaths every hour, which is in fact only a handful of deaths per hundred thousand cells. Similarly, since our original single cell ends up replicating one quadrillion times, a very rare event like a replication error in a newly made DNA, and therefore a mutation, happens constantly. An error rate of only one in one billion results in billions of mutations both in DNA itself, and in the proteins that are coded for by

DNA, another process that also commonly repeats about one billion times for each of thousands of proteins.

These numbers help to diffuse one of the most common misunderstandings about cancer: that if you get a mutation you might get cancer. The answer is that you are guaranteed to suffer billions of mutations in your lifetime. (Mutations can occur from replication errors, but also from attack upon and injury to cells, such as happens from ultraviolet light, or from triggering by viruses embedded in our DNA). A gene is a sentence with many chemical letters. If one letter of the sentence mutates only once in one million divisions, and there are a quadrillion divisions, then each one of your letters mutates one billion times!

Life would not exist if it passively accepted the gigantic number of variations imposed upon it by extremely common rare events. As we have seen, life is the imposition of order upon the tendency to dissipation. While many mutations simply die off in insignificance, and others are harmless and can be ignored, we are actually busy all the time correcting mutations in our cellular DNA. We are editors as well as writers. Cells not only divide and replicate their DNA, but they do so with higher than expected accuracy due to brilliant proteins that proofread and repair faulty DNA transcription. The proofreading has an essential bearing on the outcome of two phenomena. Remembering that our own DNA was printed out from our parent's DNA and theirs from their parents, century upon century, generation upon generation, for billions of years, we can see how replication-repairing proteins preserve life's viability across time. (Repair of transcription errors is similar to but different from the repair of event-damaged proteins, discussed in Chapter Fifty-Six).

Replication repair also helps to delay errors that lead to aging, and to thwart cancer, which usually results not from a mutation but from many of them in malevolently functional sequences. As you can now imagine, the most dangerous mutations are those that damage the proteins that repair mutations. The most insidious enemies are not those who assault you head-on, but those who infiltrate and undermine your surveillance system.

The concept of "compounded numerical gigantism" helps us to understand the inventiveness of life, which evolves over huge amounts of tinkering, variations, errors corrected, and errors preserved as new discoveries. Still, the order of magnitude of these processes requires emphasis, because it changes our total vision of life. Let's look at compounded numerical gigantism, and its implication for DNA sequences. Just how many sequences, just how much information could ever be included in or eliminated from our DNA? As we examine this question, it will help us understand the construction project in which DNA tells atoms where to go, and for which ATP provides the energy.

Chapter Sixty-Five

A Chosen and Cherished Collection

In Chapter Thirty-Three we discussed how the atoms of life are far from a random sample of the universe, and are products of selective culling. When you think about the sequences in DNA, you see in spades how life's numerical gigantism is nevertheless a small selection of the combinations and permutations which are theoretically possible. As large as are the numbers of which we consist, they are only a selection of the universe. The concept of "culled selection" applies to our atoms, our energy, and our information sequences.

DNA consists of chemical units that function like letters in an alphabet which spell out words that are the meaning of the sequence. The units are not random, but definitive. Given the billions of units in human DNA, and the choice of four alphabet letters for each unit position, we come upon this problem: If the four alphabet units were in fact randomly sequenced, you would find a branching path in which the first location could contain any one of four letters, and anyone of those four could be followed by branches to any of the four, followed by branches to any of the four. Thinking about this example of expanding permutation, as represented by DNA, which codes the proteins whose functions regulate life, Christian De Duve has written in *Vital Dust*, (see also Chapter Eight) that our genetic code, its DNA sequences, and our proteins constitute "... an extremely minute proportion of the unimaginable immense number allowed by simple statistical estimation—four to the 50th power, or one thousand billion billion billion different possible sequences for a stretch of only fifty nucleotides." In other words, if DNA sequences were random, the number of possible sequences would be vast, very vast.

But in fact the human genome contains not merely fifty but about three billion nucleotide "letters" in sequence, each one of which represents a choice from one of the four options. If you do this a billion times (creating the number four to the billionth power) you find, as Werner Loewenstein has, that the number becomes so large, that it contains more units than astrophysicists believe there are atoms in all the stars and galaxies of the universe!

This mind-smashing math reminds us of the rapid numerical acceleration of multiplied permutation. It also reminds us how much possibility exists in human DNA. If our DNA were random, it could be almost infinitely variable. The sequences available to us, the potential series of choices, is large enough to contain all that could ever exist. But we are also reminded that, though we have a trans-cosmic theoretical potential, in fact we are very limited.

We are already chosen and partly frozen. We are not a product of random DNA sequences, but of highly specific spelling of DNA "words" or genes. Our octillions are, after all, mathematically modest. To understand ourselves, we need to keep in mind the choices which are available, and the fact that a *very particular set of choices have already been made*, and placed in line billions of years ago. While our potential at the theoretical mathematical level is essentially limitless, our bodies and minds represent a long, particular, precise, delineation and limitation. We are not shotguns; we are shot out of time like arrows. We are not the blather of gossips but the aphorisms of genius. We already contain huge amounts of long-culled information, and we have already rejected much larger amounts of the possible as we saw in Chapter Three. We are variable only within a domain that is already limited and set. And yet our limited domain is incomprehensibly gigantic.

These examples from DNA reveal two things. Life is; it has been invented; it is already *an old and careful subset* of all that would have otherwise been possible. Second, the existing format of life came to be after a huge amount of experiment, trial, and *rejection* of other pathways of chemical recombinance. Life is the preservation of what worked.

In his book, *The Fifth Miracle,* Paul Davies (see also Chapters Four and Twelve), has come up with a droll manifestation of our nature as recombinations of gigantic numbers of particles. "Simple statistics reveal that your body contains about one atom of carbon from every milligram of dead organic material more than a thousand years old...you are host to a billion or so atoms that once belonged to Jesus Christ, or Julius Caesar, or the Buddha..."

This is exhilarating. I am partly both Jesus and the Buddha. I always thought so! There are so many atoms in me (our old conceptual friend, those octillions) that they can easily hold samples, numerically large samples, a billion or so, of the residues of the most divine men. But what about Hitler and Stalin? I must be partly their atomic carcasses, too? And when you finish running this thought out to its logical terminus, I am rather unimpressively populated by unwanted residues of reams of dead cats. Our personal collection of atoms may have many impressive or humble sources, distributed in many ways, based on gamblers' odds.

Christian De Duve wants us to understand cosmic life, Earth life, and ourselves, not as products of accident, but as inevitable expressions of the universe. Many DNA sequences, many chemical bonds, many macromolecules never come to be, or did indeed come into existence only to be rejected and thrown away, because life is guided by selective culling. One of De Duve's most convincing arguments rests on the power of numerical gigantism.

"Bacteria rely on large-scale genetic experimentation…A single bacterial cell would, by unrestricted exponential growth, cover the whole surface of the Earth with offspring in less than two days…" Not a bad reproductive rate. A bacteria is in the range of one millionth of a meter in size and about forty million bacteria take up residence in one gram of soil. Despite its diminutive stature, a well-fed, unmolested bacterium could soon cover all the kilometers of the Earth with progeny. A teaspoon of soil has been estimated to contain six hundred to eight hundred million bacteria of ten thousand species. What if every one of them has the ambition to cover the Earth with his or her children?

If soil numbers don't impress you, try your own body, which is estimated to host about one hundred trillion harmless, maybe even sometimes helpful bacterial guests on your skin, in your intestines, and elsewhere. (Estimates of the numbers of these commensurable bacteria, who live on and in us, vary widely but are always huge).

Once again we find ourselves in the presence of incomprehensible numbers, as the trans-countable bacteria produce their prodigious progeny. Very large numbers of bacteria could, each one, if left alone, soon cover the Earth with its children. Surely some of the ingeniousness and precision of life derives from the innumerable genetic experiments that take place every day, over billions of years. Since most of the bacterial children in fact die off, the amount of selection pressure for and against the best designer bug is gigantic. The little bacteria, who form most of the tons of biological material on Earth, are all the rare few survivors, the winners of a long race. How does it change our understanding if we stop associating "bacteria" with "disease," and instead think of them as "the best and the brightest?"

De Duve continues "...the main advantage bacteria gain from their fast multiplication rate is the enormous numbers of mutants they can offer natural selection." By the time one of our larger and more complicated human cells can divide once, a bacteria on his or her own can have produced an estimated one trillion kids and grandkids "among which several billion mutations due to replication errors alone" would have occurred. These variations produce a pool of varying choices for natural selection to cull or to champion. In spite of the fact that we primate mammals like to flatter ourselves as being the pinnacle of evolution, and this era of time being "the age of mammals," most of life, by numbers and by mass, remains bacterial. There are reported to be one hundred trillion tons of bacteria on Earth.

There are two main strategies to life. On the one hand there is the small, simple cell, with a small, simple genome, which can replicate rapidly. This is the microbial strategy. The second strategy is to create complex genes that run complex adaptive large organisms, like us. Neither strategy has ascendancy, or is more "real" or "true" to life. Neither alone is the end or goal of evolution. Both are expressions of the life force.

Life invents slowly and rapidly, in large and small scale, by trump and fiat, or over ages and eons, because it sorts and shuffles, out of the almost infinitely possible, a chosen and cherished collection, which it then preserves, yet continues to modify.

Let us emphasize the manyness with which life deals, whether it is atoms, genes, bacteria, or human generations. When we discussed cytochrome oxidase in Section IV, we were left with rhetorical questions about the evolution of such wisdom and delicacy under such dire conditions. Now we can see how the inventiveness of life derives from billions, trillions, and quadrillions of rare events. Just as life guarantees the probability of the unlikely, its motto could be: "So much happens, and so much time to choose." Life takes gigantic numbers of organisms, containing gigantic numbers of DNA sequences, over gigantic time periods, and then picks out its choices. Evolution, like the Marines, wants a few good bacteria.

We are about to use this vision of evolution to understand one of life's quintessential strokes of genius: *the quantum allocation of energy to facilitate the manufacture of us.*

We are now ready to understand our own life with wonder, based on the reduction of wild cosmic energy to quantum aliquots, that can be reliably and precisely delivered, to exact construction locations, in unthinkably gigantic numbers. Enter that hero, repeatedly referred to, long waiting in the wings, "ATP." ATP is our very own tweezers, our very own surgical laser beam, by which we tame the cosmic bull and puff energy precisely into the swirl that is us.

ATP combines the numerical gigantism of life with the precision of quantum energy. The construction of each one of us out of atoms, energy and information comes to a point called "ATP." This is the capstone.

Chapter Sixty-Six

One Memory In All Life

ATP is a moderate-sized molecule, consisting of about thirty atoms, of carbon, hydrogen, oxygen, nitrogen and phosphorus, the common stuff of life. Its core consists of a module that is also found in the repeating backbone of the gigantic, billions-fold molecule, DNA. Life is conservative, using and re-using good designs, a phenomena we refer to as path dependence (Chapter Eleven). ATP and DNA both use an early designer component that is approximately two billion years older than oxygen breathing and that has become so essential to life that all of life has gone down the path of using it until there is no turning back. This one molecular design is now intrinsic to Earth life.

On to ATP's core module has been added a unique feature, a kind of removable chemical "thumb." This thumb can be broken off, or added back, without harming the core molecular structure. The removal or re-addition of this thumb releases or uses up the exact quantum of energy on which life runs—all of life, every cell of the trans-calculable cells on Earth. The structure of this removable and restorable thumb rests on the way phosphorous atoms make chemical bonds. In the deep story of energy, as we have already seen, life is layered over the strengths and limits of chemical bonds, which in turn rest upon electron movements, which in turn rest on physics and quantum zones of the universe.

In us oxygen breathers, with our electron transport chains in mitochondria (Chapter Fifty-Four) including cytochrome oxidase, thumbless ATP gets its thumb back, and is re-energized mostly from the oxidation of food in our mitochondria (there are other minor sources). The energy rich ATP with its thumb re-attached now spreads out in large numbers among the entire cell, in every cell, like missionaries, like money, like good laws of a good government. Wherever energy is needed, energy-enriched phosphate-thumb-toting ATP goes there.

To clarify a possible confusion: we transport our energy in two currencies. When we uptake cosmic energy from the sun, plants, and food (which was the focus of Section IV) we use a subatomic coin, electrons. When we turn around and spend our energy to construct the molecules of which we consist, we use a larger energy unit, in the realm of chemistry, ATP. When we build ourselves using energy from breaking off the ATP thumb, the chemical bonds from which the energy springs also involve electrons, because change in electron states is the essence of chemical bonds, but the amount of energy involved is much more than a single electron. We earn electron pennies but we spend ATP dollars. So even as we conjure up our embodied selves from matter, energy, and information of the impersonal universe, we face the problem of inflation. Many electron journeys down our radiant scaffold are required to "load up" one ATP, and as we will soon see in Chapter Sixty-Seven, a remarkable intermediary step is involved. But it costs lots of stellar hydrogen, light waves, photosynthesis, electrons, carbon bonds, and food for us to pay the contractor who lives in our own DNA and who bills us for toenails and brain cells. That contractor freely spends ATP.

These are some of the ATP-using processes that go on inside our cells: the long library of letters, words, and sentences in DNA have to be copied, reprinted, and checked for errors; RNA has to selectively copy segments of DNA and then translate the code into living proteins; cellular components like membranes and mitochondria have to be manufactured and repaired; ions (charged atoms), have to be transported (as we saw in Chapter Fifty-Seven and will discuss again soon). Special tasks of special tissue cells have to be implemented: muscle cells contracting, nerve cells conducting or retrieving; food has to be imported; hormone messengers that communicate among us have to be heard and responded to. All of life is cellular, and ATP (or its close cousins) provides the energy throughout. For cell replication alone, a process we know that we each do about one quadrillion times, the energy must come from hordes of quadrillions of energy packets released from the removable ATP thumbs.

Estimates for the amount of ATP in a cell vary, partly due to the fact that cell types differ (for example, a contracting muscle cell in a working carpenter requires more ATP than does her hair follicle cells), and partly because such numerical gigantism is difficult to quantify in experiments. After all, a typical human cell is tiny, about one ten-thousandths of a meter. But in this tiny kingdom of life, there are estimated to be between ten million to a billion ATP molecules. These all get used, by having their phosphate thumbs broken off, and then re-energized, with their thumbs re-attached, every one to two minutes! That means each of our cells use, break, re-make, and reuse ATP billions to trillions of times a day. Since we have as many as one hundred trillion cells, we find ourselves again to be entering a realm of numbers in the order of magnitude of sex-or-septillions per day. For our ATP use for our lifetime, we again plunge into the incomprehensible octillions. ATP use by us, by life, partakes of incomprehensible numerical gigantism.

Unlike our octillions of atoms, which are units, the sex-or-septillions of ATPs that today alone, I have used, used up, and remade are a process. They are not things, with feet on the floor. They *are messengers of energy*. They are constantly in motion in space, and in transition in essence, receiving and giving energy, making or breaking phosphate-thumb bonds, moving electrons around in order to activate the bonding chemistry. The ATP is used in many cellular processes, each one of which is calling out for energy, but in the cell as a whole ATP also must have its overall proper balance, like that of a total economy, which must balance money supply with inflation or deflation. The metabolic balance of a cell is maintained in a manner which we could also analogize to our body temperature: there can be small variation within a range, but we must maintain proper body heat or die. Every one of our cells in turn must maintain its positive balance, its savings, of ATP, even though its savings are being simultaneously drained to the tune of millions per minute. Cellular ATP balances, self-maintains, self-restores, and rectifies perturbations, ceaselessly. If cells cannot maintain a balanced overall ATP budget, they die.

How can such a delicate balance be continuously monitored and maintained, at such high speeds, in such gigantic numbers in trillions of cells simultaneously in my body?

The story of ATP energy apportionment by living cells, by bacteria, apples, ants and apes is so remarkable that some people, like Harold Morowitz (see Chapter Five), speculate that energy metabolism, not DNA or RNA, is the actual origin of life. This edifice, by which I mean life on Earth, which uses cosmic energy to restore order in life's "race against decay," has been focused down onto one chemical joint, the phosphate bond on one chemical, ATP. ATP's phosphate joint is where the cosmos deals its cards, apportions aliquots of energy appropriate to and deliverable to life's myriad creations. Out of the immeasurable magnitudes of chemical possibilities that we have already seen could be coded into DNA, *life is built upon a severe constriction, a tapering, a pinpoint.* Life on Earth is a very specific creation. Life is a product of both compounded numerical gigantism and of specific, preserved minutiae. Cosmically vast potential has been honed to diamond-point uniqueness.

The ATP system is well informed about how to bring energy to life. The information for this process is passed from life to life down the eons. If this information is lost in transit through mutation, the subsequent cells are not viable.

The enormous amount of information contained in cells, in life, is predominantly generic to all of life. Cells' special identities and functions use only a small amount of information. Most of life shares with all of life a large percentage of its information. It takes almost all we have just to exist. We share with every living thing the DNA information that tells ATP what to do and how to function. Every cell in the world contains the information necessary to make and utilize this singular chemical. All of life on Earth has one memory in common, the DNA sequence that codes for ATP.

Only 7% of our genes are specific to vertebrates. All the evolutionary distance between a primitive fish and me forms only a small fraction of my genes. Our great brain, with its accompanying mind, our fancy hand with a thumb, are only details. Most of our genes are necessary for the fundamental processes of life. Not only do we share a large majority of who we are with fish and frogs and alligators, we actually share a large percentage of our genes with wheat, which has cells that need to do many of the things which our cells need to do to stay alive. How would it change our culture and our ethics if we acknowledged what we already know, that even a vegetarian eating his bread is still partly a cannibal?

Still, when we look at the long stretches of time involved in evolution, millions and billions of years, we can understand that a great deal of weaving, combining, adding, deleting are involved. Among this slow, tailored, inventive and creative change, is a countercurrent of conservatism, stability and uniformity. We share several hundred genes with even the most primitive bacteria. And we share ATP. There are two sides: evolution, everything is changing. And there is one life, and it shares ATP.

Life is an enclosed, self-propagating metabolic system that manipulates the essence of the universe and reshapes matter and molecules into itself. Before it dissipates, life grabs up a careful selection of atoms, and bends them into itself, its own long-designed molecules, by spending energy that was originally taken out of the universe, but which is now packaged in ATP.

As the study of ATP reveals, all of life shares a set of chemical construction paths that are ancient and ubiquitous, and that are activated repeatedly in gigantic numbers. From bacteria to birds, life is various and inventive, but like a culture, like a language, it is a singular inheritance. Of all the theoretically possible DNA sequences that outnumber the atoms of the galaxies, life is one antiquarian connoisseur's private collection, one ancient, artistic old ring.

Chapter Sixty-Seven

Visitations from the World's Origins

W e haven't reached the end of it yet, nor even its most wonderful secret, this story about how the explosive cosmos breathes its quanta of energy into our small cells by the millions every second.

We have already examined how oxygen breathers like us use electrons in mitochondria to get energy, but not *how* multiple units of electron energy actually get hooked *onto* the phosphate bond on ATP's thumb. This step is unimaginable. As with cytochrome oxidase, this story stuns us with the inventiveness and precision of life.

You and I are constantly getting rich from the expansion of the early universe, from the original opening, from the particles of the yawning creation, particles like electrons and hydrogen atoms. Now we are going to see how we use fourteen billion year old protons, compounds of quarks, to harness the food-derived (sun-derived, hydrogen-atom derived) energy onto our messengers, ATP. We use the cosmos's own protons to connect up our inner energy supply. In order to harness energy onto ATP, we mine the most basic building blocks of the universe.

The flow of protons over a membrane, like sheep jumping over a fence, provides the energy that actually hoists the phosphate-jointed thumb onto the waiting ATP, to make it whole and energy-rich again. Let's examine the process of proton use in detail. To do this, we have to return for a moment to Section IV's topic, electron transport for energy production in mitochondria, which will in turn tell us about protons.

Remember that hydrogen is our commonest element (Chapter Thirty-Three), occurring both in our large molecules and in our abundant body water. Remember also that during photosynthesis, chloroplasts inside green cells take electrons from water and supercharge them from sunlight to "fix" new, high-energy electrons as bridges for chemical bonds inside sugar, the origin of all our food (Chapter Forty-Six). So you can see that hydrogen's main constituents, electrons and protons, are abundant and easily separated from each other. Our personal energy comes from breaking apart chemical bonds that plants fixed in sugar (or in sugar derived complex molecules), and passing the energy stored in sugar's electrons down the "radiant scaffold" (Chapter Fifty-Three), until the electrons end in their deep trap, oxygen (Chapter Fifty-Five). But we never discussed how the electrons that fell into oxygen got re-affiliated with protons to form water at the end of respiration. Where have the protons been hiding? How are they recaptured to complete respiration and to restore the water cycle of planet life?

Unlike the swift and tiny electrons, protons, with their positive charge, are bigger, heavier, compounded of quarks, and slower. They are sub-atomic nuclear heavy weights, relatively speaking. Once abandoned by the electrons that have been stripped away by life hungry for energy, the protons move through water in our cells, softly bonding and releasing, mobile and flickering, as if lost without their better half. They move from one to another water molecule, hand-to-hand, like square dancers doing a "do-si-do," using a weaker and temporary type of bond, not the standard covalent bond. Everywhere that the negatively charged electrons move in the mitochondrial transport chain, the positively charged protons follow like sheep, like lonely boyfriends pining for their blond beauties, a metaphor that makes one wonder how closely the microcosm and macrocosm may in fact reflect each other.

At last, or should we say, incessantly, tens of millions of times per cell per minute, the fast, free electrons seduce the pining protons to jump over a membrane into a tunnel inside one protein, an enzyme, called "ATP synthase." (In Chapter Fifty-Nine we encountered the idea of tunnels inside protein folds.) Pulled by the negative electromagnetic charge on the electrons, the positively charged protons, who are visitors from the early universe, late progeny of the Big Bang, enter inside this large molecule in our cells. ATP synthase, like cytochrome oxidase, is a protein with multiple subunits, and is a hefty equivalent of five hundred thousand hydrogen atoms in size. This remarkable manifestation, this original invention, is itself a hoary presence, as old as any life on Earth. Like ATP itself, ATP synthase is found in the most primitive and the most sophisticated cells. As the electromagnetic attraction of the positive protons for their negative electrons pulls them through this great old protein complex, a piece of protein is rotated by the passing protons. A "turnstile" in this New York City subway of life rotates, and the electromagnetically driven movement of protons is transformed into the mechanical energy of a motor! All of my efforts to avoid mechanical imagery to describe life flounders here. The energy created at this rotating cellular motor is used by life, by our cells, by us, to hoist a new phosphate bonded thumb onto the used ATP, which now becomes fully charged and renewed. Electromagnetic attraction between hydrogen's electrons and protons, as contained in water, is translated into mechanical, rotational force, and, in series, the mechanical energy delivers chemical bond energy and the phosphate thumb gets reattached to ATP. (Similar protein pumps are used in other cell processes.)

Our life rests on a trick that requires skillful fielding of protons, those little old denizens of the atomic nucleus. Ancient as it is, the ATP synthase can shuttle down its alleyways three to four hundred protons per second to turn out one hundred ATP's. So you can see that it takes the energy of four electrons and their four pining protons to make one ATP "dollar." The ATP's themselves are used up and recharged approximately once a minute. One sugar molecule gives us about thirty ATP.

There is an energy chain, a sequence, that moves from the origin of the universe's early opening, right on down to the proton pumps in ATP synthase. Our personal energy comes from food, that derived from plants, which contain the endosymbiotic chloroplasts, which invented and contain chlorophyll and its related photosystems. These energy catchers use the solar photons speeding down to us from the sun's nuclear hydrogen fusion reactions which occur at unthinkable, unearthly high heat, and which release energy from ancient and original hydrogen atoms. The chain starts with the conversion of energy to matter in the early universe.

From the hydrogen atoms of the early universe which congealed in the furnace of the sun, we get our energy, and from the hydrogen atoms of the early universe held in solution as Earth water, we take our pinches of electrons and protons. The early expanding universe turned energy into matter; the hydrogen atoms formed fourteen billion years ago became both our energy, and our electrons and protons. We make our ATP by taking the broken cookie pieces of hydrogen (electrons and protons) and putting them back together again. Both our pieces of matter and our energy originate at the origins.

These impersonal, invisible, innumerable visitations from the world's origins, at last are rolled through the fingers of life. Early cells figured this out billions of years ago, by doing trial and error experiments in legions over eons. When we breathe, when we think, when we remake our energy coin, ATP, we are beholden to the early universe, to hydrogen, to water, to the evolution of ATP synthase. Our greatest scientists are merely children breaking apart and analyzing our layered, ingenious inheritance. There is as yet no human as creative as a cell.

Tactile proteins move, bend, and touch, cytochrome oxidase and its allies conducting electrons, ATP synthase threading protons, until ATP catches and carries energy on its million-fold missions to the right time and the right place in the right amount to the cellular processes of life.

I am reminded of John Milton's image in the "Sonnet on His Blindness,"

"...His state
Is kingly: thousands at his bidding speed,
And post o'er land and ocean without rest..."

Precision messaging crosses an isthmus. The messenger is ATP and the isthmus lies between mere things, and life. ATP crosses that isthmus and carries energy from chemistry into the realm of vitality. All of life on Earth is unified by recognition of ATP as the messenger, and the phosphate bond as the quantum energy unit, speeding and posting in every cell without rest.

Our wonder is not just about the inventiveness of the catalytic proteins, like ATP synthase, nor the venerability of the cosmic tinkering that the proteins embody, nor even the numerical gigantism by which they all appeared and evolved. Our wonder is about the rate. Life's speeding and posting is an unimaginable wonder.

Chapter Sixty-Eight

Skillful Motion That is Ceaseless

Life is chemistry at a very high velocity, threading together uncountable millions of messages nearly instantaneously. We saw in Chapter Fifty-Seven, discussing calmodulin, that large and relatively clumsy proteins can change their conformation in about one billionth of a second. When DNA is replicated, it adds fifty chemical letters to its length every second, and this replication occurs many times every second throughout our trillions-fold body.

What we call life, what we are, consists of very large numbers of very small units, communicating, interacting, combining, and disassociating at very high speeds, following pathways that use very large amounts of information, which guides intricate atomic placements, all of which produce self-interacting entities of octillions of units, in cohesive and mutual regulation, that occurs within infinitesimal dimensions.

In Section III we discussed "octillions of atomic placements." Now we need to upgrade our thinking. The atoms are not in place. They are being juggled, octillions of balls at the same time, and life never puts them down until the show closes after about three score and ten years.

The wonder of life resides in its frothy irreducibility. It has no stasis, no landing gear. It consists of ceaseless interactions. The wonder of life is how it consists of *scintillant combinatorialism*, speeding and posting without rest. The very essence of life is its internal galactic array of simultaneous interactions threading past each other without collision. We are predicated upon skillful motion that is ceaseless

With so many messengers, there might well be a message. With so many messengers, the message might well be quite intricate and hard to understand. Is each one of us analogous to the electrons, protons, or ATP in our body, a messenger unaware of our contextual message? Just as ATP carries a message across the isthmus between inanimate and animate, but does not have any perspective on its actions, might we also be important and meaningful in ways that we do not understand? Does my life carry an essential message, that enables the Earth, or the solar system, or the galaxy to endure, in the way that mute and thoughtless ATP molecules enable my cells to exist?

The scientific revolutions of the twentieth and twenty-first centuries have catapulted us into an expansive mythos commensurate with exploding knowledge. We now can understand ourselves as the elemental stuff of the universe, woven into infinitesimally interactional entities. We can consider our life as the transportation of matter and energy into *situations* in which they interact with information and variation to elaborate new *adaptations,* generated at rates and in numbers that create *particulate fluidity,* within coherent, temporarily regenerating *unity.*

I have somewhat arbitrarily plucked out of all the wonder of life a few emblematic phenomena, like cytochrome oxidase, ATP, or ATP synthase, to demonstrate phenomena that would also have confronted us had we examined any other life process. No matter where you might peer with your "cytoscope" into life, you would encounter the same magnitudes of numbers, rates, history, inventiveness and majesty.

Since we cannot personally experience such small sizes, such large numbers, such speed, such simultaneity, so much complexity, so much race against decay, so many mutations, so many corrections, such a hubbub and fervor of atomic and subatomic balls in the air, so carefully guided, we have no vocabulary that can fit these realities which science has revealed. The sciences of the twentieth and twenty-first centuries have led us to wordless wonder. All we share with the less informed men and women of old is that, amidst our dismay, decay, and suffering, we feel blessed.

Somehow we look out from a consciousness, and feel some love sometimes, though we are late arriving derivatives of strange essences, like the molecular structure of ATP. We now understand that we are the frothing hubbub of innumerable skillful interactions that were crafted and interwoven over billions of years. We are threaded through with *messengers and messages that connect us to the ancient, primeval, expansive opening out.* The electrons who distribute their energy to us in our mitochondria, and the protons who spin the tops in our ATP synthase, have traveled a long way to make us: fourteen billion years. For all the hair raising cosmic chaos of black holes and quasars, there are citizens like electrons and protons who follow laws.

Walt Whitman phrased it:

"I too had been struck from the float forever held in solution."

Each cell, each interaction of chemicals within a cell, each delivery of ATP energy packet, occurs in a context that is cellular, organismic, planetary and cosmic. Every motion connects two knots in a web. Things get built by messengers traveling short distances within complicated kingdoms. We may speak perfectly clearly but we do not necessarily know what we are saying. We live inside a context that is older, bigger, faster, and more numerous than is our consciousness by which we seek to understand it.

What our lives mean in the context of the universe is a problem mirrored in the context of our bodies. The magnitude, rate, and complexity of cellular chemistry does not compute in the daylight of our waking minds. The sex-or-septillions of ATP energy bonds that we make and break everyday: how are they located and routed? Exactly what organizes and directs the flickering and vast webs within us? Why isn't there more collision and chaos? How do we stay integrated and coherent for our approximately three score and ten? Why don't we implode into a molecular traffic jam, a Cross-Bronx-Expressway of chemicals? With so much high speed posting without rest, how can we possibly self-organize?

Chapter Sixty-Nine

A Self-Causing Hyperstate

As recently as the later years of Albert Einstein's life in the middle of the twentieth century, the classical dichotomy remained: either the world was understood as a field of natural law, in which every event was exactly caused by antecedent conditions, which were themselves previously caused; or the world was magical, capricious, acausal, subject to personal motives of gods and devils. As we have seen in Section I, one of the gifts of the second half of the twentieth century was insight into various types and degrees of causality (See Chapters Ten, Eleven and Twelve). Today we understand that lawful initial conditions, such as the information state at the origin of the universe (or at the origin of this expansive phase of the universe's eternal cycles) can unravel with both lawfulness and freedom, which do not contradict each other. Lawful initial conditions can spin such a web of possibility that no single outcome can ever be predicted. Events can be *simultaneously* lawful, variable, and original. Paths of cause may intersect at multiple locations, multiple times, adding, subtracting, or multiplying each other's impacts. At different phases of heat or size, causal forces may operate differently, such that relationships between fundamental particles may be different now than they once were when the universe was small, hot, and compressed. Some causal forces may scatter their impact along equations that intrinsically contain a plethora of outcomes.

Today, it is easy for us to be both scientific, and to think of the world in terms of cause and effect, rather than in terms of witches, goblins and gods, while we also see the universe as creative and evolving. Determinism, pervasively fixed causal sequences, and science, are no longer synonyms.

Of all the new insights into cause-and-effect relations, the one that will help us best to understand why we don't congeal into a traffic jam is top-down causality (Chapter Ten).

We have repeatedly encountered the concept of top-down causality. Commonly, we observe causes from the bottom up, as, for example, when we see a house being built, and the completed building seems to be a composite sum of its parts. But we have also encountered situations in which a product influences and changes its own causes, as, for example, when atmospheric oxygen, which is mostly the product of photosynthetic life, changes life on Earth, oxidizing the iron in the rocks, killing off oxygen-intolerant life-forms, and selecting for oxygen breathers (see Chapter Fifty-Five). Living things changed the planet from which they themselves were derived, changing the very conditions that initially caused them.

When we consider our own whirling matrix of octillions of atoms, none of which are exactly in place, all of which are oscillating, bending, entwining, combining, breaking apart, subject to the electromagnetic influences of their neighbors, incorporated into large molecules, or zipping through membranes as solitary ions, in the most irreducibly complex, zillion-fold weaving of non-colliding interactions, powered by the chemical bonds built from electromagnetic sunlight, we have to wonder whether the whole is merely a sum of its parts. Is it possible that the whole, once it has been formed, creates new, top-down information, that changes, informs and regulates the very parts that created it? In life, is "more" "different"(Chapter Ten)? And does the difference account for the non-collisional traffic of our froth?

We need to give thorough consideration to the problem of the origin of information inside living beings. After all, if we spin up from a fixed information chip, then we are merely machines. But if we do not have a fixed information chip, like the code of DNA, then where does our information come from? Why don't we just collapse into a mess?

To hold in wonder the soap bubble of life as it spins and shines, we need to free ourselves from simplified, mechanical, thinking.

If cells were simply coded by DNA, in the manner of software coding a computer function, then all of each person's cells, which contain the same DNA, would behave the same. The body would be the summary of the DNA code in the cells. This is true for very simple beings, like sponges, who are little more than collections of identical cells huddled together. But we have brain cells, muscle cells, kidney cells, and many other distinct cell types which all derive from the original single cell at conception. Cell type differentiation starts in the early embryo. But if this process is directed by

DNA, then what is directing the DNA to turn off part of itself, to thereby differentiate into cell types? Similarly, when a messenger molecule, like serotonin in the brain or insulin in the fat cells, sends a message to which another cell responds, how does that response occur? Every cell in my body has the same DNA, but does not operate on the same information.

The processes of cell differentiation, or cell signaling, are complicated and varied, but a core principle is involved. Based on contact with a molecule, or with the surface of another cell, a surface protein on the cell in question changes its conformation (as discussed in Chapter Fifty-Six), which changes its interaction with its own intercellular proteins, which in turn change their conformations, and a signal cascade results. This cascade, which starts with information from outside the cell, can alter the cell's essence if it reaches down to the chromosomal DNA, to block, augment, or modify how the cell will print out one or more proteins from a section of the DNA gene.

DNA is the king of the solitary cell, but in a confederation of cells, influence moves across membranes, and the chief administrator has to adapt to the interstate laws. Inside our bodies, which are always in contact also with the outside, our DNA is not always top dog. Bombardments of influences may activate or suppress some of its functions. Biological information in DNA is not a one-way street. DNA in one cell may, through signal molecules and signal cascades, influence gene expression in another cell. Mosquitoes, verbal threats, or intimate sentences may also alter your cells. Our cells function within a *context,* which can influence their information state. Cellular life not only prints out information from its DNA, but it also is a selective information absorber. Information flow is multiply directed, not linear. We have already seen in Chapter Fifty-Seven how a DNA-coded protein, calmodulin, tells calcium what to do, but in turn how the tiny calcium ion tells calmodulin what to do.

The original theories that portrayed DNA as the controlling singular source of information have been refined. Genes in DNA are now understood as being responsive to both internal (cellular) and external (environmental) cues, an emphasis known as epigenetics. As DNA lies in wait in the chromosomes in the center of the cell, the specific genes that get expressed depend upon promoter elements, chemical methylation, accessory proteins, trans-cellular factors: an increasing number of levels of feedback control. Cells perceive and integrate signals from the flux of the world. The DNA itself is packaged in a chromosome, and all the chromosomes are herded together in a nuclear envelope, and even this container, which was up until recently imagined to be only a ziploc bag, is now understood to be yet one more level of regulation over DNA implementation or suppression. The nuclear envelope has many pores, which act as selective gateways that control what messengers get into, and what messages exit from, the court of king DNA.

Genes are expressed, silenced, or tethered by checks and balances in the cell and its environment, a governance similar to the one James Madison wrote into the American constitution. Gene transcription is edited by cabinet members, press secretaries, congresswomen and many others in a process that shows both the centrality of information and the wisdom of listening. Regulation of DNA expression or suppression is a field of knowledge that is rapidly expanding with both new data and increasingly complex operational models. Informational feedback loops regulating DNA in us are probably richly woven and multiple. Even now it is being revealed that tiny sections of DNA information, interspersed in the allegedly non-coding, "junk" DNA regions, do indeed code small chips of RNA called microRNA, which in turn regulate other sections of DNA: yet one more example in the expanding set of DNA auto-regulation, self-modulation, expressive complexity. We run on self-interacting, auto-proliferative information.

The body is not a kingdom because no one ruler exists. It is not a democracy, because many phenomena are already legislated in or banned by genes. It is more like a constitutional democracy, with both limits and freedoms. Looped layers of systems attend, respond, signal and regulate. The multiple conversations produce a whole that does not resemble any one of its parts, a whole which can anticipate, proliferate, differentiate, prune back, with ingenuity and subtle refinement.

Notice two things: the whole is different from the sum of its parts; and products, like proteins, can change their very own authors and producers (top down), like DNA.

Lives like ours are leaderless entities where the totality governs. This is a process that requires very high speed, multi-channel self-communication. Life is not limited to its origins, and becomes not only more, but different as it develops, an evolving entity. Our evolution is not stepwise from fish to frog, ape to man. Our evolution is momentary and continuous, as we think, love, get sick, and decay. We are in constant reformatting mode, to some degree. (Though we are also conservative and traditional, too). We include the original information state of the universe. Inside of us "plus" still attracts "minus." But we include much more.

In life there are four billion-year-old products, which themselves, now for billions of years, have been causes. Cytochrome oxidase and ATP synthase are examples of this. DNA may be the greatest example of this. These large skillful molecules evolved slowly long ago, and for billions of years now they have been rearranging planet Earth, its atmosphere, even the composition of its rocks. These very ancient products of archaic forces now exert causal influence. They are woven and threaded into *cause and effect patterns* on a *numerically gigantic scale,* at *infinitesimal dimensions,* at *very high speed.* Within this whirlwind of becoming, there are events which are caused by molecular interactions which are billions of years removed from the worlds we call physics and chemistry. These events are products of biological systems that reproduce, sustain themselves, and form some of their own loci of cause. There are limitless interactions compounded by life influencing itself. Nor is there any final end product, since every molecule is sinuous and vital, and the living form, the cell or the person, balances between decay and renewal, adaptation or elimination: "change or die."

We are not only products of the early universe's atoms, energy, and sunlight, or the evolution coded in DNA; we are also derived from and generating right now cascades and tapestries of numerous causes and effects in high speed interactions and in complex webs of combinations and permutations, with causes becoming effects that are causes, with feedback loops of self-influencing events, with causalities in series and in parallel. The degree and kinds of law, influence, pattern and impact, escalate and magnify like the combinations and permutations of large numbers.

Thus, *life becomes multi-causal and self-causal, irreducible, not a thing.* Life is an evolving, changing dynamism, self-transforming in every nano-moment. We will keep re-examining this non-reducible, complex, self-influencing, other-influencing system of life, through several examples and metaphors, in the upcoming pages. This is the insight into life which frees us from determinism but keeps us consistent with science without believing we are puppets or machines.

Life is not merely things, energies and laws, but a *self-causing hyperstate without final definition.* Even though we were caused, we are free.

Once we exist, we gain a freedom that was not present in the conditions that caused us. We have "top-down" freedom. The freedom is a product of billions of years of interactional complexity in which causality resides at least partly within the whirling system itself. At least part of the origin of us is us. We are caused partly by the world and partly by ourselves. We communicate so well, so fast with ourselves that we are our own first influence. Even Albert Einstein actually had some choice.

We have already established that life is partly a product of its deep origins in the cosmos, which determine the laws by which electrons, protons, electromagnetism or gravity organize the world. Life is also the product of its own reams of internal interactions. Life itself, at the top of a pyramid, at the crest of a tornado of laws, causes life as we live it now. The way that our minds derive from our bodies, but gain to some degree the ability to control, influence, or even transform our bodies, through (for example) diet or exercise, similarly, life itself gains some degree of directive autonomy in shaping the evolution of life.

Each one of us is partly a product of cosmic law, partly a product of the new set of life's laws, and partly not a product at all. Each one of us is a zone of original cause, a first place, an origin. From the humming and buzzing of our zillion-fold interacting connections, unique top-down causative loci emerge. We have met the creators and one of them is us. Every mouse and man is a trial run of a new idea. The universe has many origins, most of them as yet unborn. Ten years from now there will be buds on trees in April that do not yet exist even as a single cell in today's world. More is different. A lot more is very different. The creation re-creates the origins of creation.

The "original information state" has merged with and dissolved into the information state that is emerging from its self-interactions. (Of course, the proliferation of information via self-interaction within templated information systems, though very vast, and seemingly nearly infinite to mites like us, nevertheless has some limit, and is not truly infinite. See Chapter Sixty-Five.)

Due to the exponential growth and expansion of recombinant potential, due to the numerical gigantism within life, and due to life's rich appropriation of many zones of law, life intrinsically contains, for good and ill, unpredictable new combinations of its buzzing and whirring cellular universes, combinations that partly have no origins other than themselves.

This is the biology of free will, the biology of self-responsibility. Life generates some of the complex information that is necessary to keep complex life aloft.

These abstract ideas require vivid examples, which upcoming chapters will provide. But first we need one other critical idea.

Chapter Seventy

Self-Emergence

L ife can be said to initially emerge out of the universe. Over eons, life begins to emerge from laws that emerged because life has emerged: self-emergence. Self-emergence is a concept we need to pause over. Let's try to pin it down, so that we can more fully appreciate its implications. Properly understood, self-emergence is a scientifically compatible but non-reductionistic idea that embeds life more inextricably in wonder.

The word, "emergence," actually expresses two different ideas. Obviously, new things emerge in the universe. This observation is consistent with the evolution of compounded entities, like amoebas or elephants, and does not require the radically new thinking that is associated with the idea of "emergence." More specifically, as we saw in Chapter Ten, "emergence" refers to the appearance of new laws, and not just new events. It would seem that evolution through natural selection of living beings is a fundamentally different type of emergence than the appearance or disappearance of heavenly bodies, that are born from gravity or blown apart by super-nova explosions. Natural selection requires memory; the selected organisms transmit their adaptations only if those adaptations are genetically based, held in chemical memory, and transmitted through replication of DNA. Natural selection is the emergence of a new law, in the biological zone, that does not exist for stars or planets. Stars die and then reoccur, but they do not transmit memories of new adaptations.

But two interpretations of "emergence" still remain. Emergence can mean that laws were latent, unexpressed, but implicit. The universe held these laws close and invisible, but they were always there, waiting to emerge, like the hidden cards in Texas Hold 'Em Poker. Or emergence can mean that the laws are new, brand new, totally new, not just newly expressed. (In poker, this would be cheating).

In the first view, the original information state of the universe contained implicitly everything that now exists, but merely took fourteen or so billion years to play its hand. Emergence is then understood as phased manifestation. For example, out of the genes combined from maternal and paternal germ cells, the fetus emerges. Some genes can be turned on or off, as happen in cell differentiation, to make lung cells that are different from heart cells, but the entire fetus is prefigured (it would seem) in the genotype, and cannot express what has not been coded there. My cells cannot become frogs. Maybe the universe is like that. Accordingly, in this view, the universe is a slow, pre-programmed unfolding. The universe is being born out of its mother lode of original information.

In the second view, however, the initial information state can be elaborated, combined, and permutated to produce laws that are not inconsistent with the original information state, but were never pre-figured within it. In this view, the laws of the world may be as changeable as the things of the world. As we will soon see from our discussion of antibodies, even genetics, with its apparently large but limited sequence of information, may in fact be able to generate information. Limited information may be able to amplify itself beyond its own original limits. Is the universe like that?

These thoughts about emergence bear on whether you see the world as created (passive voice) or creative (active voice). Is the universe like a lotus unfolding, beautiful, but limited to the design contained in the lotus seed or bud? Or is the world like the realm of all flowers, which appeared for the first time during the Cretaceous, more than one hundred million years ago, and which now consist of approximately two hundred and fifty thousand floral species fertilized by and responding to butterflies, bees and birds?

If the universe unfolds like a lotus, it has an origin and a plan, which it gradually expresses. All of its newness has ancestral pre-figurement. What appears new is not essentially so. If, on the other hand, the universe resembles the world's proliferating array of flowers, there is no formatted, pre-existing code. What is new emerges out of interactions, now. Every moment is fundamentally and essentially creative. New species of flowers, in new colors, receptive to new modes of fertilization, by new pollinators, will emerge, not because of any initial information state, not like a jewel in a lotus, but because the fundamentally new can emerge from the concatenation of atoms, molecules, and laws spinning in their mega-permutations, flirting with or obstructing each other, by the maha-zillions, second after second, eon after eon.

You can ask yourself: "Where was the Pauli exclusion law of electron behavior (Chapter Five), early in the Big Bang (or the early universe) before electrons even got organized into atoms (which took hundreds of thousands of years)?" Did the exclusion law preexist, in a latent state, merely unable to appear? Or did it come into being along with atoms that used electrons and protons?

Seen from the viewpoint of the science of emergence, the creation of the universe is less like a computer program written by an old man in the sky, and more like the pattern woven by fireflies on a summer night. Similarly, who could have predicted, based on full knowledge of the twenty-six letters of our alphabet, the contents of the Library of Congress?

At the core of the universe is there a message? Or are the messengers—the ATP carriers, the electrons, our own stargazing lives—co-authors? We may speak clearly, not fully appreciate what we are saying, but be able to say it better when we discuss it with our editors, readers, classmates, and teachers. The message of the universe may be less like a mission statement and more like a conversation around a campfire among friends. Creation may be both original and ongoingly contextual. Like a conversation we have with trusted friends, creation may start with a format, like the original information state, like our personal ideas, but evolve and become reorganized into unpredictable and unexpected patterns, as occurs when our friends' ideas rearrange our opinion. Is it possible that the creative process in our minds duplicates the creativity inside the universe that wove Albert Einstein, Walt Whitman, and barred owls?

Chapter Seventy-One

Antibodies and Information Proliferation

In describing features of life, I have used terms like "unimaginable" or "creative" or "remarkable manifestation," regarding skillful proteins or surprising inventiveness within life. The world does not become easier to understand as we learn more. Life and the universe continue to strike us with momentous incredulity.

On the one hand, life appears to be pinched, fixed and frozen. All Earth life uses ATP and its cousins to deliver energy around the cell. Period. Of the zillions of chemicals that could have been used, or even that could have been used as the hand to hold the energy-binding phosphate thumb, we have just one path, one choice, now operating in every cell everywhere. Life is a crusty, fixed, uniform creature, like a barnacled whale. Some things never change and apparently can't be changed, because they are held in place by an old-boys network that was billions of years in the making. So many things depend on so many other things that nothing can be shifted in an apparatus chocked full of interlocking tinker toys. Once you have built the little ships and towers of Lego land, you are no longer free to arbitrarily change any one piece.

But on the other hand, completely surprising inventions greet us at every turn, proton rotors in ATP synthase, bi-metallic catcher's gloves for oxygen, alphabets of chemicals in DNA. Life demonstrates *funneled constraints in ironic combination with essentially limitless adaptive flexibility*.

The arena in which not merely the existence of adaptive flexibility, but its degree or magnitude, became an issue was in the study of antibodies. The antibody response that we all possess raises the question of creativity within our own bodies and within the universe. Our antibody response reveals how un-planned, non-formatted, partly genetic but partly original events are coping, dueling, and inventing the universe inside us. If we want to ponder whether life is truly top-down creative and emergent, we will be edified by the process of antibody formation (first mentioned in Chapter Twelve).

As the twentieth century steamed towards its close, one of the apparently deep mysteries that continued to confront biology was how our bodies can produce such a large variety of antibody proteins which can so accurately target the infective agents who attack us. If our array of hundreds of millions of antibody types are preprogrammed by our genes, this would raise two problems. First, how can our DNA contain so much information, to code so many varieties of antibodies? After all, our DNA must also harbor enough memory to build the entire cell. Antibodies alone are just one cellular function. Could they really take up so much memory space? Also, if millions of varieties of antibody molecules were coded from memory units in our DNA, our antibodies would quickly become useless, because, as we saw earlier in Chapter Sixty-Five, bacteria mutate at a high rate, and would quickly make an end run around our historically derived, DNA-coded, old-fashioned antibodies. New bacterial mutants would get through our static antibody defenses, no matter how numerous. We would be playing football with fixed moves from last year's Super Bowl.

But if, on the other hand, our antibodies are not derived from DNA codes, where are they derived from? If they are each new mutations, we would develop few new types, because animal cells mutate more slowly than bacteria, divide much more slowly than bacteria, and have editing processes that guard *against* mutations. Bacterial DNA is much shorter and can be replicated much more quickly. Bacterial mutations would still run around and through our body's slowly mutating defenses.

The existence of our hundred-million-fold array of antibodies seemed a logical impossibility at the end of the last century, yet there they were. And our immune system, which pre-eminently includes our antibodies, is remarkable effective. (There are other features to the immune system as well, such as "cellular immunity." See Chapter Seventy-Four). While it is true that some bacteria can kill us, such as cholera or plague, as can some viruses, such as HIV or certain strains of influenza, the total efficacy of our defenses can be measured by comparing our immunity to that of a dead body. Within hours after death, a body becomes nothing more than a food supply for bacteria and other putrifiers. Even our friendly commensurable bacteria, mentioned in Chapter Sixty-Five, who share our bodies with us, do so only because of a truce between us that relies upon our continually active defenses. When we die, our billions or trillions of bacterial friends turn on us and join in the hungry dissolution of our bodies. Our admittedly temporary and flawed immunity is nevertheless spectacularly successful, and partly relies on antibody varieties that we can supply in very large numbers. Antibodies appear to defy the "central dogma" that DNA sequences are the source of our cellular information. Our DNA does not contain enough sequential information to be the source for all of our antibody proteins.

The resolution to this mystery took years of discovery by numerous investigators. The answer lies in the fact that life plucks new information from the zones of error, mutation, fracture, and chaos. Life is not merely a printout from a DNA code. Life is not merely order, but array overcoming disarray. The disarray that is being overcome paradoxically contributes to the array that subdues and partly subsumes it. Like the Romans, like the Incas, the hegemony of life expands not only through conquest, but by incorporating the skills of vanquished vassals. The proliferation of antibody types to meet new infections occurs from the melding of DNA-coded protein modules with randomness and caprice. New antibodies can be made to meet new circumstances because life is not only coded but creative.

We have already considered the fact that replication of DNA (as cells divide) is complicated and leaves room for many mutations, and that transcription of DNA into RNA, and then translation of RNA into amino acid protein sequences, are also complicated and with room for error. (Chapter Sixty-Four) In most cellular processes, error-correcting mechanisms reduce the number of mistakes. In the creation of antibody proteins, however, some errors are allowed to stand and are *reframed* as new and valuable discoveries. Pre-existing, somewhat lengthy, but essentially limited areas of DNA that code for antibody protein can be transposed, rearranged, in inventively new sequences. To create antibodies, genetic sequences that code antibody proteins are broken apart and re-arranged in a variety of ways. These breaking points form a sort of synapse, choice points, where the randomly new might become the increasingly useful.

If such spontaneous reorganizations happened to the thousands of proteins that run our cells, the newcomers would not fit in; they would not mesh with the molecular surfaces and interfaces with which they were intended to interact, and the cell in which these rearrangements occurred would die. Mutation in ordinary proteins might well be lethal. But antibodies are proteins that do not function primarily within cell systems. Their newly emergent forms are sent on surveillance to encounter viral or bacterial invaders. The rearranged antibody's crooked and misfit nature, that would render it useless as a cell component, renders it specially fit for bio-espionage and hand to hand combat with invaders. Broken and reassembled mosaics of DNA and protein sequences inform both the skill and the creative artistry of our survival through antibodies.

Rearrangements, transpositions, shufflings, and inversions of informational sequences create new information. Arising from gibberish, new antibodies accost invaders with renewed efficacy. The world and its living organisms are simultaneously coded, chaotic, and inventive. Antibodies are proteins that are coded by constantly new informational sequences that arise in cells from pre-existing sequences that are re-ordered by an array of capricious and rejuvenating processes. The huge numbers of continuously new antibody types express the potential of combinations and permutations of limited numbers of genetic sequences.

In spite of our adulation for Einstein, we can now all confidently contradict that revered and shaggy sage. His playful term for the cosmos, "The Old One," not only plays dice, but plays Pig Latin, and like us, creates Italian out of Latin, Hindi out of Sanskrit, and speaks an evolving living language that can express new things in new ways. The world has pre-coded laws, as do cells. But laws also allow for flexible combinations and woven patterns that permit the utterly unpredictable.

Randomness constrained to certain zones, but randomness none-the-less, is paradoxically one form of cosmic lawfulness. Cosmic law, biological laws, use pools of randomness as pools of opportunity and invention (as well as storehouses of decay and entropy.) As part of the lawful nature of the universe, bounded or regional unlawfulness is one of the laws.

Great fanfare accompanied the first sequencing of the entire human genome, and rightly so, as it was a stunning step forward in human awareness. An animal had become self-aware, knowledgeable of microscopic cells, informed about bio-informatics, and able to isolate and study DNA. We had also penetrated into the alphabetical code of our own life, and spelled out its master sequence. But sequencing the human genome has also led to disappointment. So much remains unknown. Probably the fixed linear information contained in the given DNA strand is only the start, and other life processes, similar to those which create antibodies, rest upon information skimmed from DNA regulation, combination, rearrangements and mosaics, by which karma can capitalize upon the potential fertility of constrained chaos. Feedback regulation of DNA by RNA is probably involved as well. We have also considered how the information of the total system may be more than the information that initially coded the system, due to self-interactive feedback (Chapters Sixty-Nine and Seventy).

In living systems, information can proliferate. Antibody-producing cells can generate genetic information that uses but is not limited to the information in the genome. There are more actual football plays than all the quarterback's have called in all of history's football games.

The story about antibodies has deepened our understanding of how evolution operates at the molecular level. When we think of genes as malleable molecules we understand them differently than when we think of them as units. They are not printed out exclusively as templates from fixed sequences. Time and circumstances play upon molecular genetics in DNA the way a jazz pianist plays upon the keys. There are sequences, recognizable patterns that play and replay. But as the patterns replay, mutations create subtle divergence between the first expression of the melody and its semi-repeats. As more divergences appear and accumulate, there is an expansion of the gene pool, like an expansion of melodic potential. Genetics, like jazz, means pattern, variation, and rearrangement. Out of the melody of an "old chestnut," the jazzman plays his improvisation. Out of sequences on the DNA, antibodies form from rearrangements and shufflings and transpositions.

In this world where we remain wonder-struck children, delighted, curious, and not-fully comprehending, there is a lot more music than there are notes. There is more music than can ever be sequenced. The world's most dour pessimists have never worried that we will finally run out of musical sequences. Just as the pianist is not required to stick to a single melody, but can add them, antibody-forming white blood cells select genetic sequences from the cornucopia in DNA, and then mutate, re-arrange, combine and reverse them to form improvised antibody protein sequences that preserve some modular structures, like the melody, while selectively expanding them into new patchwork creations.

Proliferate, make errors with aplomb, and then keep what actually worked: the genetics of antibodies mimic the life of any creative writer.

Chapter Seventy-Two

The Music of Life

U nderstanding the invisible molecular realm of genetics helps us to understand our large-scale waking world. By more carefully examining genetic laws, we can more clearly see life's ability to amend the constitution of the universe and to write its own statutes. Genetic molecular processes explain three features of life's capacity for variation and adaptation, helping us to understand why there are lobsters and whales, mathematicians and musicians.

First, some of what happens to antibodies also applies to the DNA code in its entirety *throughout the history of life*. Many of the processes by which antibody information exceeds the information of static sequences—rearrangements and the like—also influence the entire history of the genome. Across time and history, across billions of years, the genes of living beings follow the processes of variation and divergence. However, proteins within a cell, which form an old-boy network, cannot mutate as wildly as the roving, freewheeling antibodies, but the general processes are similar. In the history of DNA evolution, we find fixed, invariant melodies of favorite old songs, such as the codes for ATP. We find subtle melodic shifts that make only nuanced change (such as the codes for ATP's close relative GTP). We find parallel overlap and redundancy, where a slight variation occurs but mostly preserves the old song, so that the old and the new gene are similar. This helps explain *conservation within the context of change*, and why my personal DNA shares some tunes with wheat and gorillas. We have diverged, but we have partially preserved and overlapped. The new genes have varied, but only subtly, slightly, and the old core melodies remain recognizable.

Viewed against the backdrop of billions of years of life on Earth, molecular genetics provides us with a model of how the universe creates. Genetics reveals balanced governance by conservatives and liberals. There is preservation that is constitution-like: for example, every cell has DNA and RNA holding information, or ATP carrying energy. And there is innovation: I do think of myself as different from wheat and hopefully even from a gorilla. The world evolves with path-dependent rigidity and with antibody creativity. What we see on the surfaces is mirrored also in genes. Our human gene pool remains woefully similar to our experimental mirror, the mouse. Experiments on mice have some relevance to human medicine because our gene pools overlap to such a large degree. Then again, the few differences that in fact do exist between mice and men are big ones.

A second feature explained by "genetic jazz" is cell differentiation within the embryo, as we just discussed in Chapter Sixty-Nine. Just as antibodies can play with genetic sequence, so gene expression in a cardiac cell as opposed to a neuron, is probably partly dependent on transpositions and other forms of rearranging gene sequences. (As we are about to see, other forms of gene regulation are also involved). Cell differentiation shows how the same melody can be expressed in different ways at different times and in different places.

A third feature of the jazz-like property of molecular genetics has to do with gene expression, which is the summary term, for antibody formation, cell differentiation, and now all other forms of gene expression. Gene expression of all kinds is a more fluid, creative, jazz-like process than is generally appreciated when we say that something is "genetic." The term "genetic" has become falsely associated with the concept, "fixed." (See Chapter Sixty-Nine.)

After giving so much emphasis to numerical gigantism, and then inflating it even more in Chapter Sixty-Four to "compounded numerical gigantism," how embarrassing to smack head on into this confession: there are thought to be only about thirty-thousand protein coding genes in our human genome. (This is a very approximate number). When there has been so much emphasis on complexity and quadrillions and octillions, and when one-billion-to-one has been stated to be a good bet, how can a man and a woman derive from slightly more parts than a Saab or an i-Mac?

"Ah, yes...." compounding, transposing, combinations and permutations. Sequencing the genome, as great a step as it was, has provided limited knowledge, because genetic expression of DNA is enhanced by "promoters," and shut down by "reducers." Genes are regulated, managed. There are some cell biologists who believe that, rather than there being a master pattern, each gene has a unique regulatory system. And who is regulating the expression of genes within DNA? Our old friends. First, DNA creates proteins that regulate DNA, in feedback and top-down patterns. Proteins can regulate their own origins and the origins of other proteins in complex, regulatory loops. And, second, the environment, other cells, the whole organism, can regulate the expression of each gene in DNA, as we considered in Section IV regarding protein conformational allosterics. Environmental molecules provide information to and influence other molecules in their environment. (Little calcium ions modulate the protein, "calmodulin" and vice versa). Genetic control is networked. The thirty thousand proteins that our human genes code for can be turned on and off, fully or partly, rapidly or slowly, and in varying sequence. Each one influences many others who also influence many others. Our thirty thousand or so proteins are not only the building blocks that make us. They are also active participants in a cart-wheeling, exquisitely timed circus act. They are few in number but many in function. Our "building blocks" after all are not blocks, but the living, bendable, smart, sensitive proteins we have already witnessed in Section IV. Rather than meaning "fixed" or "predictable," the term, "genetic" should be used to refer to one instrument within an orchestra, or one frigate within an armada.

Once again, we find that life does not print out information in the manner of a photocopier. It expresses information, with pattern, variety, even rhythmic pulses. One theory about differentiation of cell-types in the developing fetus is that everything depends on *timing*. We become human not only due to our information code in DNA, but also due to exactly *when* we promote its thirty thousand famous old melodies. Our genome is less like a computer program and more like a sonata. It is our essence to be pulsatile, rhythmic, and in the beat.

The play of pattern, divergence and even syncopation that makes for jazz, or for an Indian raga, may well represent a human discovery within music of one of the essential features of the universe. When we play jazz or a raga, we are congruent with, and mirroring, a cosmic process. The core cosmic information state seems to be such a terse, physics-based Law Book. That's all that seems to have existed in the Big Bang. Yet it has emerged into our world as the crowded hubbub of toucans and tanagers flying through the rain forests of Costa Rica. It has emerged as novelists inventing plots.

Amplification of simplicity into variety due to sequencing, timing and compounding—that play of law and caprice, of karma and chaos—this is the secret of life, the secret of cosmic evolution, the secret that Einstein missed when he dichotomized mathematical law from the roll of dice. The "Old One" not only plays dice, but He probably listens to Keith Jarrett's Köln concert and to Ravi Shankar. We are all simple melodies gaining inflection at the hands of an old (billions of years old) master musician.

The toucan was not invented by a formula. The information source for an organism, for the toucan, for me, is the whole organism, which responds to its own inner code through networks, cascades, and loops, and which responds to the biosphere with which it is in continuous contact. *There is no one place to turn for information, origins, or laws*. That most ludicrous and impressive frugivorous appendage, the toucan's bill, is like a melody in a long improvisation piece that has lasted for billions of years. Who needs more than thirty thousand melodies to choose from when you expect to be playing their notes and resequencing them over billions of years? This musical image is one more way to capture the meaning of the concept of emergence. A very large, very old, very numerous, self-influencing system rewrites its composition book as it whirls on. The system has become complicated enough to keep reinventing itself.

The wonder of it all is that electrons, the Pauli exclusion principle, hydrogen atoms, photosynthesis, mitochondria, cytochrome oxidase, calmodulin, and ATP are participating in my writing this sentence, but none of them have a clue how to do so or what it means.

We can awaken our sense of wonder by seeing our own life in the same way. History, contexts, patterns, woven and rewoven over eons older than stones: wonder is simply the habit of acknowledgement. We know enough to be rich with wonder. We cannot box our world of toucans, jazz, and novels into any one narrative.

Chapter Seventy-Three

Law and Freedom

The creative universe is like a writer. Or, it might be a better sentence to say: writing (or speech) mirrors the way deeper forces work.

The chemicals in fixed sequence which we call a "gene" is like a word, that can be combined into various sentences but cannot be used anywhere. The gene "word" adds meaning only when coherently embedded within a bigger sequence (a sentence). That sentence in turn is a contributor within the context of a paragraph. The same words can be written in different sentences, which then make complicated paragraphs that emphasize or mute particular words, incorporating or excluding them from what gets finally expressed. The genetic world, like language, is essentially combinatorial, a mosaic. And some words are just left out, lingering in the dictionary to await their day. But the linguistic meanings available to us are enormous and situationally responsive.

Words may be limited, and genes may limit us. A mouse lacking the genes for higher thought can never obtain a B.A. from Princeton. But the organism can modify its expression (to some degree) in response to changes in both internal and external environment. We speak at different times and in different ways.

Our personal information state is partly limited by our genes, partly an inventive recombinant expression of our genes, and partly held in reserve outside of our skin in the great library of the universe. We are most quintessentially human when we are aware that we are not just ourselves. We combine words into sentences that are new and we speak and listen to those who are around us. Our information state is not entirely knuckled within us. Words, speech, connect us, bring us into interactions. So, too, genetic expression is not a forced march, but a dialogue.

Once again we find that individual life cannot be understood as an expression of a code of DNA, of genetics, or of any fixed and sequential information system. A whole cell is always a product of a whole cell. DNA in isolation never codes for life. Even viruses that consist of little more than a strand of DNA (or RNA) must parasitize a cell and use its system-wholeness to generate new life.

All life expresses previous life. Every cell exists within the history of life, and within a biosphere of other contemporary lives. Life can activate DNA to kindle information back into life, but a strand of DNA is not life. It is always contextually (historically, cellularly, biospherically) expressed and regulated. Life exists in connections, not in isolated components.

Even if you told me that tomorrow a scientist at the Rockefeller Institute will take a single strand of DNA, put it in his future Nobel Prize winning test tube, and print out a complete, breathing mouse, that strand of DNA would still have to be understood within the context of its history (where did the scientist get it from?) and its context, including modern civilization, Rockefeller University, research labs, and that particular scientist's mind. The strand of DNA could only exist in the first place because other mice of the past evolved its information, and could only express itself in a proper, effective context (the mythical scientist's lab).

When we forget life's embeddedness, we see living *things*. When we remember, we recognize that every life is embedded within the *vortex of being*, a primeval swirling, and inter-messaging upwelling. Within this updraft, no one thing exists. Every being is entrained by the history of its genes, by the chemistry of its environment, and by the intrinsic living process. Even the term "gene" is a newspaper-article-level simplification, because the so-called sequence varies from person to person (genetic variation) and is always functionally integrated with regulator genes, or with non-coding sequences that may nevertheless influence how DNA bends and folds and therefore how that "gene" gets expressed or repressed. For every "gene," there are many regulators, which often regulate how other regulators regulate, in tailored and idiosyncratic systems. Every being exists in the context of its ancestor-derived genetics, in the context of its internal, proliferative, emergent information, and in context of its environment, which expands outward without clear limits. Every boundary or edge is only a pause. Every "thing" is the repetition of a memory that is itself temporary, a kind of stutter.

No one can predict my next sentence.

Chapter Seventy-Four

The Self

To feel the wonder of life we have thrown away superstition and belief, and then we have jettisoned linear causality, reductionism, and the theory of single-cause origins. We have watched the world become new things in new ways, and like children, like bloodhounds, our minds have followed into paths of discovery. There has been no resting place. Every location has had its context and led us to new connections. Every moment had its antecedents and subsequents. Energy, atoms and information move. If everything is motile and boundless, then what about me?

Antibodies open up the question about "self" and "not self." If everything is connected, continuous and in conversation, then why do I feel separate, unique, my "self"? Antibody creation looks like a form of war that seems to define us. What is our sense of "self" and where does it come from? Why do we need to protect our biological "self"? Why don't we just merge, blend, and go with the flow of the universe? If life is so combinatorial and contextual, then why do "individuals" work so hard to remain clearly bounded? Look at how much energy antibody formation demands of us.

Each one of us ceaselessly proliferates antibodies. The molecules pour out of groups of cells cloned by the cell that originally shuffled old DNA and thereby invented that particular new antibody. The antibody-making clones may be stimulated into greater productivity if their products are useful, or may decline and implode if their inventions fail to serve. The process of antibody creativity, like the process of creating new bacteria by rapid mutation, works only if numerical gigantism is involved both for antibodies and for cells that clone them. A phenomenon called "hypermutation" may lead to a fine-tuning of the antibody to fit the enemy it is attacking. When it comes to the creation of new antibodies, each of us is a madcap inventor and a reckless gambler. Estimates of the actual numbers of antibody designs that we can make vary widely, from hundreds of millions to billions, trillions, or quadrillions. The enumeration of antibody manufacture quickly overwhelms our sense of proportion. About one hundred billion antibody-making neutrophil white blood cells enter our blood every day. Some cells can secrete antibodies at the rate of two thousand molecules per second. On any day you are sick with a "flu," you may secret trillions of antibody protein molecules. Compared to any one of us on a sick day, the war of the worlds looks like a skirmish. When you have the flu, you become a fabulous inventor of weapons to protect yourself. Each one of us is like Star-Wars.

Antibodies are only one part of the immune system, which is multilayered, using cells, and many other kinds of cell-made molecules, like interferon, cytokines and others. This ferocious defense of "self" began with life but has existed in its current sophistication for about five hundred million years, since genetic rearrangement and clonal antibody production first began in fish. The full immune system is a very complicated, dense mesh of defense that has both redundant and alternative pathways of invader-destruction. In this context we should remember that pathogens are of many species and have their own molecular tricks. There are still five hundred million cases of malaria every year leading to a million deaths. HIV has killed twenty-seven million people. Then again, billions of people, and uncountable other mammals, fish, and such, roam the Earth alive and well, due to our ancient and modern, special-fit recognition, detection and defense immune systems. If you are alive, it's a product of defense. You and I are kingdoms at war.

While the immune system cannot represent the world in hypothetical space, as our minds can (Chapter Six), it can recognize and respond to the texture of "self" and "other" felt in three dimensions. In some ways, immunity is a facet of consciousness. We are well prepared to meet our unmaker. We are very vigilant to the interface between "self" and "other." The immune system feels this difference by touching the shapes of alien molecules. Immunity is non-representational, three-dimensional hyper-awareness, an advanced aspect of *knowing*, *separation* and *difference*, at the subtlest molecular level. We finger, feel, and fight the alien, through specially attuned molecular sensors. We design and employ a stunning array of antibody *surfaces* that expose bad *fits* (See Chapter Fifty-Nine).

Still, we should remember that out of the cosmos-transcending number of permutations that are available to sequential bio-molecules like antibodies, only billions, trillions or quadrillions are chosen for the immune task. That's a big number but a small percentage. Of the other zillions of sequences, that are theoretically possible, as discussed above in Chapter Sixty-Five, most are just too weird to do the job. Our work of self-separation selects from the universe a specific collection of antibody tools. We can make a lot of antibodies, but far from an infinitude of them. Though we create new aspects of ourselves, like antibodies, we cannot create capriciously, because a lot of our "self" is fixed.

While the zillion-fold biosphere is proliferating pathogenic bacteria and virus in unending arrays, and while we respond creatively with our own inventive battalions of antibodies, our permutations and combinations nevertheless can hit their limits. But the zillions of pathogens have greater variety. In a universe compounded out of large numbers of tiny things, the unexpected is absolutely certain. For us to survive for one day, we have to compound, permute, invent, like a sword fight. For every living thing every moment, in our biosphere of bugs, the message is: create or die. Every cell is a fountainhead of creativity that is stimulated by the need for self-defense.

Within the constraints of the past, new life is born to identify and preserve itself against the world from which it emerged. Historically preserved genetic memory, coupled to recombinant inventiveness in each new being, has a force, a life force that works to stay alive. Self-recognition, other-rejection, and immunity express this force.

Now we have come at last to a final biological conundrum. Life is a series of messengers that speed and post over land and ocean without rest, like ATP, like cytochrome oxidase or calmodulin, all of which communicate and relay atoms, electrons, protons, or electromagnetic energy. But life is also a battle, with an arsenal, and a network of defense, in which antibodies and their back-up immune systems beat back the hordes of putrefying little lives, and in which organisms with one hundred trillion cells cohere in a bounded alliance against everything else. Life encircles itself within belligerently guarded self-definition. Christian De Duve's proliferating bacteria could, but fail, to cover Earth with progeny in a few days (Chapter Sixty-Five), because someone else is already in the way. For bacteria, Earth is a zone of turf wars. We all compete for space and resources.

Bacteria would probably be pleased if we inoculated them with the genes of real estate lawyers. They all are out to get more property.

More than 99% of all species that have ever lived are now eliminated…that is 99% of species that have lost 100% of their individuals. There have been an estimated fifty billion species. Is life an upwelling, an entity, or a battle line? Are we primarily communicating, or hurtling invective? How did it come about that individuals need to defend themselves against the very bio-cosmos of which they are a part and on which they are co-dependent? Our "self" is both a temporary sensation and an assertive surge. So much of this book has been an effort to emblazon the unity and interconnectedness of life, and in spite of that, we have descended into a den of antagonistic survivalists.

We all communicate with copper from old suns, electrons of the earliest days, ATP of life's Euro-like, shared currency. We all spin, hover, and juggle octillions of atoms in the ceaseless froth without rest that is us. While we are always interactive, receiving sunlight, electrons or atoms from the world, and giving back carbon dioxide, water, science and poetry, we are also bounded, guarded and separated. We are both continuities and discontinuities, interactions and nodes.

The world is not smooth but lumpy. Space has its nodes of galaxies. Galaxies have their nodes of stars. Life has its lumps, bacteria and us, sequestered within membranes and skins. Life rests upon loci of density, where matter, energy and information can compound and temporarily increase communicative interaction. Life is lived in buzzing huddles. Generic, plain-vanilla cosmic space does not support life. We separate ourselves temporarily in order to give back more ultimately. *We emphasize separation to enhance interconnection.*

What we call "life" is temporarily bounded and jealously guarded *interactional densities.* The temporary densities of matter that we call stars give off their radiant signals because an enormous mass of hydrogen held together in one place by gravity contributes something different to the universe than hydrogen floating far and free without density or proximity. Similarly, biological lives for a moment cohere and convey. The density and separation that we contain are catalytic.

We are messengers without a fixed code. Like antibodies, we contain something genuinely original and new. We guard our personal, individual existence in order to parlay onward increasingly complex messages. We do so with a time limit, just like a star.

Every apparent individual is a temporary swirl that uses *identity, assertion,* and *destruction* of others to maintain our own *interlocking, self-perpetuating system* that is in fact always *porous* and *communicative,* always exchanging with the larger energy, matter, and information systems in which we are imbedded. We are localizations and nodes, but we are not the entities we feel ourselves to be. Deterioration and decay are as much a part of us as are our defenses, boundaries and skins. We defend only for a time densities where the *potential for creation is optimized.*

We cohere for a while, and nature promotes that coherence, with antibodies and the like, in order to maximize contact, connection, interaction, communication, and combination. We generate mutually influencing *interactional proximities.* We create immune defenses to preserve our blistering concatenation, our frenzied yet orchestrated whirring, our octillion-fold humming of atoms speeding high on energy in swirls and loops held in place by infinitesimal trajectories swerving past each other, everything a messenger, speeding on matter and energy in scintillant arrays. That is us: high intensity communicative interaction.

Without temporary selves, interactional compounding from density and proximity could not occur. Even more importantly, bounded loci of becoming, by which I mean myself, you, and cockroaches, are essential for storing and transmitting memory. Life defends itself against other lives to preserve the adaptive sequences of DNA that it holds within itself. To define itself, DNAs make self-markers for all their cells, so that antibodies can quickly detect cells that are either "self" or "not-self."

If you look upon it from the perspective of one life, there is a struggle for survival, and antibodies are weapons in the struggle of evolution through antagonistic competition. If you expand your vision, and look upon it from the perspective of Earth life, or cosmic life, immunity resembles a debate rather than warfare. Molecular hordes, such as bacterial cells, or such as people, jockey for position, true, but the entire process continues the communication of molecule to molecule, cell to cell, photons, electrons, protons, atoms and waves of energy, intersecting, combining, interacting, building, destroying creating two hundred and fifty-thousand kinds of flowers, creating Albert Einstein and frogs. The destruction occurs in service of pruning back to allow more adaptive and seasoned flowering. All the struggle and conflict is an advanced biochemical dialectic. The temporary entities are not just froth and bubbles, but information transmitters. The biosphere is like a biochemistry department where everything is competing for tenure.

We have a sense of self because we have skin and because we have DNA. We are messengers carrying big briefcases, or laptop files. From the past to the future, every life is a packet of skills and advice. We are temporary, but we aren't arbitrary. Our desire to speak and to be heard is a psychological analogue to the biological message which every life strives to transmit.

For humans, ironically, part of our message is to be receptive, attentive and filled with wonder. We separate our insubstantial selves from the universe for a while in order to carry our message, our jazz, our novel, our science, our wonder, across a little unit of time and space. We are messengers bearing a signet ring. We are always crossing the river. Our "self" is memory and information messaging to tomorrow.

Chapter Seventy-Five

A Vision of Life

Wonder is a form of courage, because we all have tendencies to dismiss whatever is puzzling, and to believe whatever is socially sanctioned, soothing, and acceptable.

Wonder stretches our credence while halting before gullibility. Wonder includes courageous perseverance in the attempt to understand, while holding at arm's length the superficial relief that might come from credulity.

John Milton, whose poem on his blindness envisioned the world analogously to the way we envision cells today (Chapter Sixty-Seven), with thousands of messengers speeding and posting without rest, concluded this poem with the belief that faithful attendance upon the wonder of the whole is more important than action. Milton used the word "wait" in its old sense of active attention, as when the servant "waits" upon the king. He wrote:

"They also serve who only stand and wait."

Here we see a reverence for awareness as much as for action. What we apprehend may be as important as what we do about it.

We can now hold up for wonder a vision of life that includes its many atoms, its multilayered information drawn from every zone, its minute spaces like "the serviceable nook," its high rates, like proteins that fold in billionths of seconds; its compounded numerical gigantism found in the numbers of atoms involved, the rates of bacterial proliferation, the permutations and combinations theoretically possible to DNA, the ATP turnover in one body, or mutation rates across a whole population; its ceaseless fluidity, so that even protein structures are snaky and allosteric, and so that no part is static or still; its jeweled and skillful sensitivity to the smallest of things, like copper ions or electrons; its jazz-like ability to compound and amplify information, as antibodies do; and its communicative, networked, interactive information compounding, so that, within a cell, proteins re-shape other proteins, proteins and DNA reshape each other, tiny ions like calcium tell proteins like calmodulin which way to fold, and photosynthesizers and oxygen breathers form vast biospheric, responsive, mutual interdependencies.

Only when we let go of our parochial sense of numbers and proportions can we relate to the *magnitudes, rates,* and *timescales* by which the simple laws of things have been compounded into the hyperstate fluidity and complexity of life. Only when we let go of the argument between "law" and "chance" can we grasp that we are children of an ancientness that looks upon the sun and Earth as newborn transients; and that we are writing science and poetry that the cosmos is hearing for the first time.

Not that the cosmos needs us, with so many galaxies, so many planets, so many authors around. We are only a part, and the sum is more than all of us, and is different. The hardest part of wonder is the courage to let go of conviction and conclusion. We can't even guess how different all of that "more" is. In exchange for all of this letting go, we get to ride our planet, and our galaxy through *the strangely fertile spaces of the nearly empty universe* from which we have sprung, and to gaze out at all the passing scenery with the contented dazzlement of wonder.

We are trying to understand why the world sometimes breaks in on us with wonder. One way of answering this question is that we have been put together in a way that exceeds both our cognitive capacities and any fixed description. If not knowledgeably, then at least intuitively, we encounter a world that is always too old to comprehend, too many to grasp, too complex to specify, and brand new. We ourselves, our own living processes, are filled with wonder and we see it reflected also in our babies, on beaches, or in the stars.

We are as skillful and clueless as ATP, while we rearrange the Old One's electrons and atoms and communicate the messages of the universe. We are its buds, its fingers, its thoughts. Something is happening, but we don't necessarily know what it is. Beyond anything we ever say, we are communicating something. In wonder, we have suddenly heard our own echo.

SECTION VI

A Constitution of Wonder

Chapter Seventy-Six

Revelations and Revolutions

Sections I through V have referred to many recently discovered ideas. Of course, wonder is not limited to those new ideas, or to any one idea, or to any set of them. Socrates and other ancients felt wonder, and it is to Socrates that we attribute the aphorism: "wisdom begins in wonder." The Renaissance, and the eighteenth and nineteenth centuries had scientists and artists of wonder. But around the twentieth tentury something historically unique happened, and we are the inheritors of this great dispensation. Over the last hundred years or so there have been not one, but scores of shattering, revelatory new paradigms that have shifted our fundamental sense of reality. Out of this very large array of new understanding, a new consciousness has been born.

Wonder is no longer the prerogative of the brilliant scientist, nor the marginal prophet. The cascade of paradigm shifts of the twentieth century has led to widespread wonder, and to liberation from many beliefs, dogmas, and conclusions. Heightened wonder has also decreased (in some minds) the violence provoked by ideas that require buttressing via threat rather than reason, evidence, or investigation. The dozens of paradigm shifts piled up on each other form a meta-shift, or meta-paradigm, across all arts and sciences, of which wonder is the inevitable and essential component. The meta-shift is not about more, newer, or surprising facts, but about a mood, an attitude, an expectation called wonder. We have come to expect revelations and revolutions in our perceptions about the universe and ourselves.

Section VI will list twenty paradigm shifts of the last one hundred years, more or less (some of which overlap with the late nineteenth and early twenty-first centuries). Each of these has already been referred to in Sections I through V. Section VI can be read as a crystallization of the ideas referred to in the preceding ones. The order presented here vaguely follows history, but many of the new paradigms took decades to evolve, many grew out of well-established nineteenth century ideas, many evolved in parallel with other ideas, and some are more easily understood in light of their followers and explicators. All twenty serve as an explanation of how wonder grew from rare seams hidden beneath beliefs, conformity, and mind-control, into a broadly embraced orientation for tens of millions of people.

There have been many previous psychological revolutions that followed empirical insights. One might think of Copernicus, or of Darwin and Wallace, who removed Earth and humanity from the center of narcissistic myopia. These men and their ideas were great moments of psychological history. My focus on the twentieth century and its near penumbra has to do with the thundering cascade of insights, and how much further they pushed us. Those great geniuses like Copernicus and Darwin paddled hard to move us downstream. But eventually we were swept downstream in a flood.

The number, twenty, is obviously arbitrary. Many of the paradigm shifts I include here might be seen as a compound of several others. Some of them might be understood as parts of a previous one, and differ only in emphasis. Some are associated with an individual; others derived from dispersed exploration by many people. All of them established a radically new psychological reality. All interlock with the others inside an embracing psychic shift. All have filtered into the waking consciousness of the average informed adult. All are mentioned in our basic education. All pop up from time to time in *New York Times* science columns. Section VI lists these twenty paradigm shifts in a way that makes them accessible, explicit, resting side by side, expressed briefly and colloquially enough so that they can all be contained in a single overarching picture of a brave new world that can no longer be censored by orthodoxy, nor accepted as blasé, nor reduced to machinery. Each of these paradigm shifts has splashed newly colored realizations that can't be whitewashed.

We already exist after the apocalypse. Through the series of observations, breakthroughs, and insights listed in this chapter, we have been liberated to some degree from fear, entrenchment, or fantasy. We are citizens in a global human community that validates wonder as an idea, emotion, experience, and an orientation.

The fact that the community of wonder has members everywhere does not imply that we are in ascendance. Nowhere today is fully free of the twin tyrannies of infantile narcissistic fantasy, and of hate-filled insistence on a local superstition. In fact, there is nowhere that is free of button-pushing, one-sentence thinking, sound bites, snide quotes, dismissive mockery, obsessional distraction, or frantic joinerism. Another reason to try to create a brief, containable frame for the new world of wonder is to post a billboard, to stand available for people who may be trapped behind a fence but who are peering hopefully over it.

Our explosion of new realizations has also created a vision that is too grand and multifaceted to easily hold in mind. No one can claim to understand or to speak for all we now see in the world around us. We are too encyclopedic a culture to produce true "Renaissance men." Therefore, a conversational version of our world-view is justified by the need to sometimes keep the whole within a single viewfinder. While the details of many new discoveries require professional expertise to understand, the total picture harnesses our abilities to conceptualize, integrate, and appreciate an interlocking series of "Ah hahs!"

This long-enough and short-enough list of the paradigm-shifts that installed the modern world of wonder is like a fractal in which every part holds the whole. Most of the great insights are codependent on others. It is like a portrait by which we can stay in contact with a charismatic face. Although this list can be faulted either as over-inclusive, or as foreshortened and neglectful, I think we deserve a touchstone.

This list is an attempt to write a constitution of wonder. These are the twenty principles through which we now understand ourselves and our world.

Our psychological world has opened out into an unending series of creative new mornings. The topic is not science but psychology: how we see, feel, experience our world. The list is not about scientific discoveries, but about their psychological implications. We are looking at how new information and ideas reconstructed our world with wonder at its center. We are inheritors of moments when the constrictions of ignorance and belief were eliminated by contrary evidence, moments when our psychological worldview opened into more unrestricted and inconclusive questions and exploration.

A startling sequence of revolutions has made us (through cultural evolution) almost a new species. We possess information that exceeds our metaphorical and narrative capacities, surrounding us with numinous newness. Psychologically speaking, we inhabit a new world that has no envelope, no dominant narrative, no container. But our new world is not one of merely relative narratives and of cynicism. We are animated by perpetual expectation.

If you keep in mind all of the following constitutional articles, you will easily be granted a lifetime passport to wonder.

Chapter Seventy-Seven

Twenty Luminous Jurisdictions

I

The Cosmos

The cosmos has incomprehensibly large dimensions.

The cosmic revolutionary was Edwin Hubble. He was the discoverer and spokesperson for the existence of other galaxies, for the billion-fold expansion of our sense of space and time, and for the uncanny recognition that the universe is an explosion. While Copernicus's intelligence and courage make him a properly revered icon, Hubble's discoveries exceeded those of Copernicus the way a tyrannosaurus exceeds a bacterium. After Hubble, the most fundamental questions about who we are have come to rest upon our location within an ungraspable expanse. While Copernicus, and Darwin and Wallace, challenged the Earth-centered, human centered, daydream, Hubble permanently erased it. We cannot now find our origins, purposes, or meanings in any intellectual construction that ignores the great caverns and canyons of inter-galactic space.

It is notable that Hubble is somewhat less famous than he ought to be if he is in fact responsible for the greatest of all human scientific and spiritual revolutions. Hubble himself harnessed tenacity, perseverance, even belligerence in pursuit of a new reality that many people already intuited, hated and feared. He was a master of careful observation and rigorous math. He eschewed rhetoric and waited until he had collected a royal flush of facts to lay out on the table. But the dimming of his stature has to do with the inevitability of his discoveries.

As the technologies of telescopes expanded, what Hubble worked so hard to unveil became visible to ordinary planetarium goers. The telescope named after him now snaps detailed photos of the galaxies he proved existed but could barely see. A very large array of sci-fi writing makes the

Hubble cosmos its baseline assumption. Edwin Hubble is dimmed by the shattering conclusiveness of what lay behind the curtain that he himself lifted. The universe is bigger, older, and has more locations (galaxies, pulsars, quasars, black holes, stars) than could have or can be imagined. To understand the Hubble cosmos, we have to surrender to an unthinkable world. The number of galaxies, stars, and planets are too large and too far apart to comprehend. Our sun, or our planet are small punctuation marks in an almost infinite world.

The psychological expansion of time and space that occurred with the discovery of galaxies is possibly the single greatest revolution ever to occur in human thought. Because of its implications, most people continue to deny it.

II

Unification

Matter and energy, space and time, the human mind
and the world it studies, show unities.

Einstein and Hubble can be understood as complementary. In terms of cultural trajectory, Einstein's reputation expands. He remains enthroned as the emblematic man of the twentieth century and as the subject of a large intellectual industry. He proved that time and space are properties of the world and are not abstract coordinates within which the world moves. He proved that matter and energy are inter-convertible forms of each other. Einstein drew a picture of a universe that has "thingness," an entity, within which everything is held, even the gauzy, apparently insubstantial things like time.

Hubble's cosmic revelations rest on light, the vehicle by which his telescopes received information and his eyes saw. Einstein, in contrast, was a creature of thought. The exhilaration and adoration that accompany Einstein's name and picture are partly due to the triumph of mind that he represents. He understood what he could not touch or see. His mind parsed the universe's mysteries. During the very same activities by which he diminished our cosmic self-centeredness, he demonstrated a profound congruence between human thought and cosmic realities. His math, his thoughts, were homologous to their cosmic origins. He came from the universe and his mind could travel inside it. No one has ever done more to emblemize the power of the human mind. We all can decipher the world and think thoughts that echo from its depths, so Einstein tells us. His name became synonymous with brain, mind, and cosmic vision.

The psychological paradigm shifts of Einstein's many discoveries also include a new position for the humanity of the scientist. He understood the world not only with his mind, but with his compassionate passion for causes, portrayed repeatedly in photos of his basset eyes. Einstein proved that a modern thinker, a physicist and a mathematician without a lab, who was a slow, remote, deep cogitator, could become a prophet. He showed us the world as a bending, warping time-space totality, and he made us feel ennobled, because our hearts and brains express aspects of the world we revere. We revere Einstein's reverence. He showed that cynicism is not a stigmata of intelligence. Einstein showed us that the human mind and heart can understand the invisible from which math and reverence both flow. The solution to all equations is reverence and wonder. His Holy Grail of a unified field theory has turned out not to be a formula but an emotion.

Einstein's ideas are among the most difficult to truly understand, but they circle around a monism that unites: matter and energy are one; the world is one with time and space; the human mind is one with the world it scrutinizes; the heart and mind contact the same deep truth; science and cosmic religious feeling have the same final goal.

III

Expansion

The universe that we can contact is dynamic, expanding,
a process in transformation.

Although the idea of "The Big Bang" was implicit in Edwin Hubble's discovery that the lights in the universe were all moving away from each other and the universe was therefore expanding, a retrospective extrapolation was required to formulate a universe that had once been small, hot and compressed (See Chapter Thirteen). One implication of this idea is that ancient India had insights of uncanny discernment. No other culture had postulated similar dimensions and time scales in its mythologies. More important to most people today, however, is their personal location within a dynamic, living universe, that in some ways resembles life, an organism. A second psychological implication of a small, hot universe is the question of origins, how the thing got going. And the third implication of the Big Bang is about endings: how far and wide will the expansion be and what comes next?

Previously, the universe was understood through human time dimensions that could be measured by generations and narratives that contained human-like creators, motives, and causes, and by a static, external "eternity." Today our sense of the expanding universe is senseless. How can an Earthling make sense out of a universe 14 billion years old and billions of trillions of miles wide? It does not connect to person, motive, cause, origin or end. Instead of something, or nothing, we have transformation.

Our concept of origin is now motile. For example, since we have seen in Chapter Thirteen that oscillation and vibration seem to occur in the smallest quantum zones, at the edge between something and nothing, maybe such fluctuations, caused by nothing, spontaneously fluctuated into something, a random, acausal quantum theory of origins, associated with Richard Feynman, Stephen Hawking and others. The origin is itself a function of change. Change is the first principle.

There is general agreement that Edwin Hubble was correct but too restrained. The universe is not only much bigger, but its rate of expansion is also faster than Hubble predicted. This escalating expansion was discovered in the 1990's when a particular type of supernova, which has a relatively fixed brightness (based upon the Chandrasekhar limit, see Chapter Nineteen) was discovered repeatedly to be too dim, and therefore too distant, to exist within the supposed confines of the universe! The universe was found to be bigger than had been thought possible. Two competing teams of Americans later shared the 2011 Nobel Prize in physics for discovering accelerated cosmic expansion, a phenomenon which contradicts the prediction that gravity would eventually break expansion with attraction. Possibly "dark energy's" expansive pull overpowers gravity. The universe not only appears to be an explosion, it also seems to be gaining momentum, an accelerating explosion. The rate of change is changing. The origin was change and the expansion keeps changing.

The implications of endless change and expansion includes multiverses, emerging out of the infinite possibilities contained within infinity. Steinhardt and Turok, (Chapter Thirteen) as we have seen, believe there is no beginning or end, but an endless series of bangs and contractions, and they base their thoughts on the added dimensions that string theory imputes to reality. There are no edges or endpoints, only further arenas of flux.

All of these suggestions fly wide of science, data, and investigation. How do you exit from our universe to see if there are infinite others? These theories are science-derived speculations, untestable, and unsatisfyingly abstract. The idea of an expanding universe and its origins and ends leave most people who think about it feeling an uncertainty that is more uncertain than the false certainty of uncertainty theory. We live in a matrix of cause and dimensions that we know we can't really swallow.

The expanding universe provides an explanation for entropy, the dispersal of everything. But the expanding universe itself remains unaccounted for. The densities, complexities and creativities within us, our own Earth-lives, provide intuitive ballast (though not powerful argument) against the defeatist attitude that the universe is "purposeless" and "meaningless." Despite the universe's opaque origin, goal, or ungraspable magnitude, we feel that something of equal salience has been organized in our bodies, minds, and planet. Nihilistic intuitions are too conclusive for the dynamism we encounter.

The expanding universe places us in a position of maximum intellectual modesty and maximum religious feeling. Every night we look out into a cosmic bed on which we sleep, an ocean on which we float, a magnitude across which we voyage. The universe is now a ponderable that is imponderable. The Big Bang in our minds leaves us floating like airborne seeds, waiting for some explanation like rain on desert flowers. We gaze out at dimensions of exhilarating incomprehensibility. We are clueless and we can't help but wonder.

IV

Aggregation

Our senses misinform us about the world, in which
material aggregates coalesce out of pulses.

The "is"-ness of matter is an illusion. Matter is a condensation, a creation, a compound of constituents increasingly elusive and subtle. Our palpable world, our gravity-grounded feet, our strong and delicate palms, our bodies and their surroundings in the hard and bric-a-brac world, are not the permanence that we generally take them to be. Matter has not always existed, and matter is not in fact compounded out of small units of itself as the Greeks believed about atoms and elements. Matter's grains, when sliced smaller and smaller, change their names and their essence. We and our waking worlds are glued together out of droplets to which we cannot impute any of our experiences. Our senses misinform us about a world that actually rests on internally layered, hastily evaporating pulses. Matter emerged out of energy during the fiery early universe through the aggregation of mass, charge, spin and other properties that seem too abstract to account for the Earth or for a baseball. Physicists imagine that those properties emerge from even smaller and more fleeting particles.

When particle physicists slam minute transient things together in colliders such as SLAC at Stanford, The Fermi National Accelerator in Illinois, or CERN's Large Hadron Collider near Geneva, they study how the high energies and heat during the early universe became matter, and their experiments and formulas take them into a realm beyond the human sensorium. Their scientific journey is a victory of intense, collaborative and competitive intellect, and also the invention of arcane, mathematical narratives that do not immediately shift our sense of reality, except when we pause over them and allow them to totally undermine us. The particle physicists' world is hidden inside the compounded world that we know. Its objects never run around free as even tiny electrons do. The particles in the Hadron Collider (for example) do not do things that ever influence us, except to make the world.

Today we understand that there are layers, chambers, strata and dimensions that fill the universe from within. The holding-together of things as they are now may be one phase, or one history, of the universe's strings, quarks, bosons, muons, neutrinos, and other waves, particles and proto-things. Particle physicists like Sheldon Lee Glashow (Chapter Ten) view all particles as products of "fields." Forces and energies are the matrix in which the material world resides. The coherence that platforms us may be a law, or a quirk in the history of our, or all universes. We cannot casually assume that the world we think we know is an accurate portrait of the world in all of its potentials.

Inside of everything is an aggregation, a joining, a connection between parts that can neither be understood yet nor taken for granted. It is still not clear whether matter was inevitable, or just happened. The world of mass and matter is an expression of deeper realities at the level of energies, vibrations, and oscillations. Much smaller than atoms, the components of things are not comprehensible through matter-energy dualism, cannot be contacted by ordinary perception and consciousness, and are intangible and unthinkable except through particle accelerators and mathematics. The known world arises from, and can lead down into, bridges that cross over from nothing into something.

V

Combinatorialism

In selected patterns and unlimited numbers,
the material world facilitates combinations in layers and webs.

There is a rush to combine. The invisible and inanimate things of the impersonal world seem eagerly affiliative. Matter not only has the capacity to combine, but a tendency to do so. As deep as you go into the nature of matter, you find yourself at a juncture where things have joined. We have come to understand that the world we inhabit is quintessentially combinatorial. This worldview gained momentum in the early nineteenth century, when John Dalton founded modern chemistry with his theory of atoms of exact size that combine in fixed *proportions*, to form all the molecules of the world. This was much more precise and revealing than the previous Greek hunch about atoms. It became an even deeper insight due to both twentieth century particle physics and chemistry.

Everything around us is a compound, and the compounding repeats and is deep. The very unfinished investigations of particle physics nevertheless convince us that even properties or attributes of matter may arise from combinations of temporary, unstable particles. The preponderance of matter over anti-matter, and therefore the very existence of the universe, might be due to asymmetrical oscillations of neutral B-mesons trillions of times in a second. Mass itself may have been added to the recipe of the world by Higgs' bosons. Various flavors of quarks join to form protons. In the "standard model" of particle physics, the world is a composite of "quarks," "leptons," and "force-carrying particles." Protons, neutrons, and electrons are the subatomic components of atoms. All of these particles, which are named for the ghostly entities discerned by mathematics and experiments, lie buried deep inside our tangible and variously textured world. But we do not encounter them alone. Combinatorialism is a force across layered dimensions: sub-sub atomic, to subatomic, to atomic, to molecular, to cellular, to organismic, to biospheric. The world we encounter is always a product of invisible things that are constrained by laws to cooperate in creation.

At the subatomic level, the vibrations of mesons or the aggregation of quarks do not collide and pile up. Combinatorialism is orchestrated. Protons have distinct properties and are not just quark omelets. Electrons obey exclusion laws and avoid traffic accidents even in large, multi-electron atoms. The universe is not just small, temporary, hasty pulses and particles in combination, but in increasingly solid and *reliable* combination. The particles are floating, but the laws that guide them have cosmic, archetypical power.

At the atomic level and above, there appears to be an urgency of combinatorialism that is both forceful and permissive. Atoms form molecules in a panoply of ways that manifests as the variety of things in our palpable domain. Chemical bonds are built by electromagnetic plus and minus attractions, but the details of these joints are unendingly inventive, leading to our cornucopia world. Combinatorialism builds a slower and more solid world, and also one open to inventive recombination.

There is not merely a coming-together, but a lattice of directives about how to do so. We are built up by layers of substances and properties that guide and augment *further* lawful and reliable interactions, and this in turn permits the addition of new layers of matter and energy. The pressure to combine contains directives that makes some combinations probable (like quarks to protons), some combinations possible (like neutrons locked in an atomic nucleus), some improbable (like a rare element), and many impossible. We and our world are products of what DeDuve has called "the combinatorial imperative." But we are also products of combinatorial exclusion, by which laws have organized our world and outlawed alternatives. Hence, Lewis Thomas' surprise that we exist at all. There is nothing built up out of "minus attracting minus." The combinatorial world derives from a federation of laws, and all *other* worlds have been exiled and outlawed.

So we need to refine our thinking to say that some things combine in some ways in vast and immeasurably complex networks. In selected patterns and in unlimited numbers everything is layered and webbed. The universe and our bodies are networks of things and properties in relationship.

We have to wonder whether our biosphere, our society, our member-ships, and our marriages, are following some general adhesive principle that also guides electrons and quarks. We also would be wise to remember that the combinatorial directives are distinct from both randomness and also from nothingness. It is a source of wonder that the combinatorialism out from which everything extends is constitutional.

If there were in fact a multiverse, and some weird monkeys from an-other universe zoomed into ours, and captured me, and held me hostage until I could explain what was going on in this universe, I would buy my freedom by saying: "Although everything is running away from every-thing else, things combine."

VI

Skepticism with Readiness

Black holes and other counterintuitive discoveries demand
that our thinking balance skeptical rejection of unsupported claims,
with readiness to embrace expanded evidence and exemplary thought.

Everything we know so far is limited, and certain to be overturned. The universe may well contain properties and substances of which we are as yet unaware. A classic reminder of this occurred when nineteen-year old Subrahmanyan Chandrasekhar, with a merely provincial, partial education, explained to one of his mentors at Cambridge University, Sir Arthur Stanley Eddington, who was one of the most revered scientists in the world, why stars over a certain size might develop so much gravita-tional compression that when their fuel ran low, they might either collapse into a density that vanished, or explode (as we discussed in Chapter Nine-teen). Eddington, a pacifist Quaker and a self-proclaimed mystic, publicly humiliated the young and vulnerable Chandrasekhar, chastising him that physics needed common sense as well as mathematical logic. Eddington could not tolerate a mathematical conclusion—and he was a brilliant mathematician—that demonstrated part of the universe could implode and disappear.

Today expressions of the Chandrasekhar limit, like black holes, su-pernovae, and neutron stars are part of elementary school curriculum. Quasars are thought to be regions of the distant early universe that rotate around large black holes. "Dark matter" is believed to account for much of the gravitational phenomena of the universe, while "dark energy" may help to explain cosmic expansion. Dark matter and energy may be dark because they do not interact in any way with electromagnetic photons, light. Matter itself may be the residue of a duel with antimatter, and the universe around us may be only a fraction of matter left over after most of it disappeared in the antecedent clash. Invisible neutrinos and anti-matter may flicker around us beyond our awareness.

A clash between matter and antimatter at the origin of time and space? Fusillades of neutrinos passing through our bodies all day every day?

We can no longer be complacent, arrogant, or conclusive. We expect ongoing revelations.

Just as our senses misled us and do not show us the realms of particle physics with their quarks, bosons, muons, or B-mesons, our light world, our electromagnetic world, all of our experiences and even intuitions may be partial, merely animal adaptations, and fundamentally misleading. Deeper cosmic realities may be invisible and counterintuitive, as black holes were to the great Eddington (who was open-minded enough to eagerly embrace the esoterics of Einstein's general relativity).

Credibility is belief extended without adequate challenge, testing, or intelligent skepticism. Wonder is a readiness that never shuts down. Chandrasekhar and Eddington remained devoted and mutually admiring friends. Wonder and skeptical rigor can clash, but they do not rigidify into entrenchment.

On the one hand our psychological world of wonder is based on evidence and information, and on the other hand we have learned to relinquish personal experience in favor of culturally expanded evidence and exemplary thought. We have to be prepared to both reject and accept. People who wonder do not hold on tight. There are both errors and revelations in both dismissals and beliefs. We need to be skeptical, flexible, and ready for more.

VII

Natural Continuities

Our personal energy and matter derive from the origin of the universe,
and have been reformatted in stars, whose light we live on,
and whose elements we contain.

Eddington misconducted himself with snide skepticism toward Chandrasekhar's counter-intuitive mathematical proofs, and by not extending skepticism to his own mystical convictions. In the popular world, Eddington remains famous for purportedly having said about the universe, "Someone, we don't know who, is doing something, we don't know what."

Among many other reasons, Eddington is revered for his mathematically (yes) rigorous assertion that stars must include outward radiation pressure. Eddington asserted that the gravitational mutual attraction inward of all the matter in a star has to be balanced by the outward force of radiation pressure from nuclear fusion reactions in the suns' core. Nearly half a century later, Hans Bethe (who was also a leader of the Manhattan Project and one of the inventors of the Hydrogen Bomb, and who later became an anti-nuclear activist) and his colleagues, were awarded the Nobel Prize in Physics for proof that thermonuclear reactions, involving nucleosynthesis of helium atoms from hydrogen's protons, were the source of energy and light in stars. They confirmed Eddington's insights.

These deductions and proofs provided the link between sunlight, whose children we are, and the nuclei of primeval atoms, whose children we must try to understand we also are. That stars shine due to radiation issuing from nuclear fusion was the flash of insight about how stars could emit so much light (not by fire!). At last humanity had the golden link in the chain that connects our personal existence, which eats sunlight to build our molecules, with the energy packed into electrons, protons, and hydrogen atoms during the cooling of the early universe. Before Eddington and Bethe and their colleagues, we were orphans, who imagined we had been adopted by the gods. After them, our pedigree directly connected us to stars, hydrogen and the origin of matter, energy, and time.

When stars shed their mystic luminosity and were understood to be products of more wondrous cosmic law, a series of Nobel Prizes went to William Fowler and others who showed how the hundred-plus elements of the world were forged in stars or supernovae. Mendeleyev's nineteenth century atomic chart of chemical elements, with their orderly progression in size and number of protons and electrons, now could be explained as a lawful creation. Astrophysics created all the balls in chemistry. The world inside of us, the world of carbon and other chemicals, was not created on day one of the Big Bang, but was built up over billions of years. If we are a thought in the universal mind, then it is a slow, careful thought, maybe a musing, or maybe a carefully worked out formula.

We are starlight. We are atomic. The elements we contain are original energy alchemized in solar gestation. Not only our energy, but our atoms, are stellar. Aside from hydrogen and helium, the atoms of our planet, like carbon, phosphorous, iron, or uranium, were made inside stars or their wild children, the supernovae.

We now inhabit a psychological world in which we understand ourselves as products, as features of the universe and its processes. We have emerged from natural continuities.

VIII

Transience/Dispersal

All material phenomena, even the supposedly eternal stars,
are transient, mixing order with dispersal, just as we ourselves do.

The eternal heavens aren't. We are the first generations to live in a dynamic universe. Humankind has always imagined that stars are eternal, symbolic of the enduring transcendent—but not anymore, not to us. Stars are born due to the clumping together of hydrogen due to gravity, or due to gravity and other forces such as uneven densities within the early universe's inflation, or due to contributions made by dark matter. But stars are created, born, and, even more provocatively, they evolve, starting with varying sizes and densities, depleting their fuel at different rates; and they die. Twentieth century astrophysics, from Eddintgon, Chandrasekhar, to Bethe, and right up to the twenty-first century moment, has unleashed many and sometimes conflicting narratives about how star-death stabilizes into white dwarfs that use up little fuel and glow on, or that collapse into black holes, or that blow up as supernovae and spread the elements they manufactured out into space, to recongeal as second generation stars, or as solar systems with planets made of stellar ash, like our Earth. There are probably many types of supernovae, a phenomenon that you can't study directly and that rests on distant brief observation (lasting less than the hundreds of millions of years ideally required), and on computer modeling of complex uncertainties such as the dynamics of turbulent flow.

The stars are now in space, not in heaven, and our planets are like sparks from a campfire. We fly through space on a cinder. The moon is not a goddess but is part of us, some lost material, our own leftovers. The solar system is new and doomed. Our universe is ruled by change. The stars are not in a heaven exempt from change and death. Nor are they exempt from our study and insight. We live in a psychological worldview in which no thing, no material, is fixed, steady, free of the master wand of entropy, change, dispersal. This is profoundly disquieting. We are the first large group of beings for whom transience is the central fact. When we look up into the night sky, we see our own origin, our end, our vulnerability, our motility, our incomprehensibility.

But the radiance has not gone. The galaxies and stars, those nuclear fusion furnaces of the cosmos, still ignite a resonance with something in our being. Creation and destruction are continuities, and we ourselves are glowing with energy waves from the nuclear lights. Starlight continues to glow inside us. We not only find stars beautiful, haunting, and inspiring, but we are their offspring. We are transient children of bigger impermanence.

IX

Uncertainty

The universe contains domains that are intrinsically uncertain,
as well as domains where laws and uncertainties mingle.

Because the universe is built over a transformation of energy into matter, and because at its deepest layer matter is oscillating with the arising and passing of sub-sub-atomic particles, and because particles may also be waves of energy, there is a depth at which change and uncertainty, arising and passing become inescapable. You cannot pin down the butterfly wings of neutral B-mesons, neutrinos, or even our more familiar electrons. Motion, change and uncertainty cannot be eliminated. These uncertainties not only rule in the evaporating and miniscule quantum domains, but they reach up into our large-scale world. As atomic isotopes, like uranium, decay, they seem to do so in a manner that permits a lawful prediction of their overall rate of decay, but that makes it impossible to predict the behavior of any single atom. Isotope decay contains intrinsic uncertainty.

Nevertheless, uncertainty has been over-generalized by popular science. Our computers may sometimes crash, but it would be hard to argue that we cannot predict and control electron behavior to a large degree. Uncertainty is domain specific, and collapses into our familiar, law-bound world in which all basketballs fall down towards the center of the Earth. The motto of the universe must be: "variation within bounds." There are realms of certainty and of uncertainty. There is variation but not caprice.

We have been fast-forwarded into a psychological world-view in which we can recognize both aspects of reality. There are laws that give us high reliability in our predictions, and there are less definable domains of every size. The weather and the stock market and our final fatal illness all partake of the universe's duality: some degree of order, some predictability, some inescapable uncertainty. The giants of the universe, too, like supernovae and black holes, follow some rules, but also elude prediction. Super-heated, sub-atomic plasma roils in turbulence across unthinkably large light-years during unstable supernovae explosions. No actuary will sell you life insurance in the midst of one of those things.

We do not need to personalize the world, nor do we need to project human motives onto its variability. We feel and harness some reliability and order. We can learn to eat healthily and drive safely. Maybe we can also learn to die by letting go of all our atoms as deciduous trees do with their October leaves. Certainty and uncertainty are shades of gray. The universe is neither random nor a steam engine. Certainty versus uncertainty is a false anthrocentric dichotomy, inaccurately plastered over the complex relationships within the world.

In our current state of wonder, we stand like jugglers with two bowling pins spinning from our hands into the air. We recognize order without imagining that it issues from commands of a motivated being, and we recognize uncertainty and creativity as properties of the same constitutionalism.

X

We are Chemicals

We are combinations of chemicals
put together by the laws of the universe;
we contain so many chemicals of such complexity and variety,
and so many laws, that we become self-modifying and indeterminate.

We are chemicals.

In my personal hierarchy of paradigm shifts that have put wonder at the center of our lives, first comes the cosmos of billions of galaxies, and next comes the recognition that we are chemicals. That we are a satchel of atoms and molecules is an insight that leads to a personal psychological revolution, requiring us to recognize the laws of the universe live in us and we emerge out of them. This revolution also reminds us how complicated and numinous the laws of the universe can get. I think and feel with molecular arrays. The laws of the universe do not just predict information about rocks rotating in rings, like Newton revealed about planets. The realization that life emerges from chemistry places wonder inside me. Emerging out of the filigree of complex chemical networks, here I stand with my robust thoughts.

The greatest triumph of chemistry is biology, its complicated and creative subdomain. The calcium in our bones not only comes from milk (or broccoli, or tofu), but it originally came from the anvil of the stars. Milk or tofu may be the vehicle that carried it around, but calcium was forged in the astro-furnaces. We are cosmic chemistry sets.

Ironically, our "chemical-ness" simultaneously tells us that we are "things," as well as liberates us from "thing-ness." We are "things" because we are a very large array of atoms, like sand washed up on the beach. We have been built grain by grain in the art studio of the world.

But we are not "things" because in the interactions, networks, connections, oscillations, resonances, and relationships among uncountable numbers of immeasurably swift exchanges, "more is different," and "very fast and fluid" is categorically different. Out of the para-infinite options, contained in mega-combinations, over eons, the flexibility of chemical combinations has spun up a creative, indeterminate, self-modifying becoming. Chemistry weaves life that exceeds "just some more chemistry," in the same way that a Van Gogh's *Starry Night* painting exceeds oily colors on cloth.

XI

Chemical Pathways

Our chemicals follow an intricate, historically derived set of paths,
like cellularity, biofeedback, homeostasis, metabolism,
and information storage and replication.

We are not merely chemicals, but chemical pathways. We are built upon a network of ancient chemical roads. We are biochemistry that, after billions of years, has figured things out, worked out the shortest routes with the most parsimony and resilience. We are the current events in the history of biochemistry. We are a throbbing ancient capital with the old thoroughfares still carrying throngs of traffic. We are the specific chemical pathways remembered and replicated by life. We not only follow laws, but historically fixed, well-rehearsed practices that have been discovered and learned by ancient cells.

Patterned and reliable chemical paths orchestrate what our chemicals do. Our internal roadmap is too complex for us to understand right now, but there are a set of rules-of-the-road guiding many sub-rules. Our chemical paths require enclosures, like homes in a city. Cells are complex chemicals that are contained, self-interacting, self-propagating, and always dependent upon and communicating with chemicals, nutrients, signals, and information outside themselves. Inside these homes, domestic rules are paramount; there is biofeedback, in which events produce effects whose outcome influences the ongoing event. There is some degree of homeostasis, in which biofeedback is harnessed to maintain temperature, energy, shape, in relative constancy. (Our temperature never ascends to 120 degrees F nor descends to 70 degrees F.) There is metabolism, which means taking energy from outside—food—and rearranging the atoms and energy into more of oneself, in skillful, rapid sequence. Here is where the prince, ATP, organizes the realm. There is information stored in three-dimensional bonds of chemicals, preeminently but not exclusively the library of life's past adaptations, DNA. There is even scheduled cell death, in which cells make way for the next generations, and also avoid unwanted replications that might lead to cancer.

We are chemicals obedient to systems that are very, very old. There are so many of these channels, sequences, circuits, and feedback loops that we don't know them all, but we know that we are cities whose foundations lie in cosmic antiquity. We are memory, karma, the sequelae of antecedents, beholden to a Magna Carta predating the British one by about four billion years.

XII

Information: A Governance of Marigolds

The universe contains information about what must happen, might happen, or can never happen, and these rules can interact, elaborate, recognize and relate to each other, enabling both coherence and complexity.

Equally mysterious as the existence and expansion of the universe is its lawful matrix guiding the combinatorial web of becoming. Rules seem to have emerged with the universe. There are facilitated *channels* like electromagnetism, and *barriers*, like the limit to the number of electrons and protons that can join to form an atom. (No atoms contain many more than one hundred protons or electrons). The origin of the universe's information is in the Dhamma, as the Buddha called it, or in "the Mind of God" as Einstein called it. The existence of information is coincident with matter and energy. The origin of the universe's information is unlikely to be understandable to us until we assert the improbable claim that we know all laws and their particular origins, in spite of the fact that there are forms of matter and energy that we only dimly apprehend, such as dark matter and energy, antimatter, mass-producing muons that are so fleeting and embedded in other particles, or other emerging scientific phenomena. We are far from being able to describe the origin of information.

Just as the ancient civilizations imagined the world consisted of four basic substances—earth, air, fire and water—we also have our creative quaternary: matter, energy, entropy, and information.

Our recognition that the way-things-are expresses cosmic information constitutes a revolution in consciousness. Everything is held inside some basic statements; everything is situated within a basin where rules have shaped the possibilities. For all of the disarray, like the athlete dying young of leukemia, there are doors that cannot open. Like a well-designed constitution, there are basic principles that also permit nuanced applications.

The laws that birthed and expanded with the universe can also converge and cooperate. A marigold startles us because it brings together electrons, protons, atoms, electromagnetism, photosynthesis, ATP, DNA and so much more into a temporary coherence. Within the cosmos that is mostly black, cold, and empty, there is also a governance of marigolds.

The matter of the world is intrinsically combinatorial, and the laws of the world coexist and interact. Potentially antithetical twin directions, complexity and entropy, cohere in all the compounds that form the world.

The information, the legislation, can cooperate, compound, and create, and at a particular crossroads, like a marigold, we can recognize, as we peer inward through the layers, the handicraft of relationships among the imperatives and barriers that we call laws. The statutes of the universe pass smoothly alongside each other. They have a mutual background. They work together, sliding easily among each other's luminous jurisdictions.

XIII

Interactional Densities

Individuals are temporary, uniquely formulated, interactional densities of cosmic materials, energies, and laws, so both transiency and universality are valid viewpoints from which to view our personal embeddedness and emergence.

Within the smooth and spacious continuities of the world there are nodes of matter, energy and information. As stars clump out of thinly sprinkled space, organisms form interactional densities within the dispersing universe's tides.

To understand ourselves, we need to appreciate the temporary format and the timeless principle.

"Beings" self-maintain, integrate and regulate materials and processes for a period of time, creating the visual kaleidoscope of tigers, toucans, and citizens. These nodes, these beings, are held aloft within webs of large cycles of energy and matter. The large cycles include the creation of molecular oxygen by plants and the use of oxygen by respirating organisms. The larger cycles swing far out of our planet's domain to include our major energy source, the sun, and the materials of our solar system or our galaxy.

The nodal beings within these great systems also compact and compound information. For example, my personal energy supply is distributed by ATP, which requires phosphate, which is leached into biological systems from Earth rocks, which came from ancient suns, where atoms were created along the lines of subatomic physics, in which only certain kinds of combinations of electrons, protons and neutrons can occur, which in turn rests on properties of electrons such as charge and spin. Thus, my personal energy supply is a product of cascades of laws compounded and woven over eons of space and time, but then compacted into a local density called—"me."

I am a compaction and interaction of every law. Every marigold is a cosmic crossroads. All laws are drawn down and woven into temporary, semi-permeably separated, layered, self-interacting systems of high magnitude and speed called a "being." Every person is a mirror in which cosmic processes are reflected. Every being is a pocket of the universe where a unique combinatorial potential has been actualized. Our thoughts involve the motions of chemicals and require the movement of the universe's tiny rocks and rays. We are embedded and emergent. We have no separate existence. We are frothing hyperstate compounds where octillions of atoms skillfully zing past each other following fourteen-billion-year-old physics laws, four-billion-year-old biology laws, and freshly minted variations.

XIV

Earthlings

Along with cosmic law, we are partly products of the specific circumstances and precise historical paths of events on Earth, which makes our existence dependent on certain local conditions.

We are Earthlings. We have a special connection to our planet. Earth is our platform and its history is our local owners' manual. We are partial products of its approximately four billion year old conditions.

For generations we have been aware that our distance from the sun and our atmosphere form essential prerequisites to our temperature-based chemistry. We have more recently integrated into our consciousness many path-dependent peculiarities of our Earth platform. Earth's chemical composition provides our substance. Earth's size determines its self-compression, internal heat, and layers of molten and solid. The motion of the upper layer over the dynamics of the inner laters has driven continents in plate tectonics, which has probably both facilitated and hindered life. Vast extinctions of life may have occurred from the dramas of continental drift (and also from batting practice in our region of the solar system with its asteroid impacts).

The experiment of life on Earth wove cosmic principles with local happenstance and eons of experimentation.

We remember in our cellular DNA big Earth-based discoveries such as how to use the energy of electrons stolen from other molecules' chemical bonds; or how to safely dispose of used-up electrons in oxygen molecules. These discoveries were made here, in our own labs. Trees and flowers are relatively new yet ancient (about one hundred million years old) and paved the biological backroads and byways for us.

Fossils, mirrors in rocks, give us one narrative about our long ancestry, and comparative DNA sequencing gives us another way to trace time, relationships, and divergence, and these two stories dovetail well.

We are special creations, but we do not come with an L. L. Bean guarantee. That we are products of planetary conditions and history is our most sobering and elegiac discovery. What a long and wild Norse saga before we got a part! Earth is our author and no character can defy the pen of the creator. The scripture about the wonder of life is written by oceans, atmosphere, and the babies of bacteria, plants, and other lives. The drama of floating, crashing, and tearing tectonic plates over billions of years has been catalytic and cataclysmic. There may be cycles of balance and relative stasis intrinsic to planetary litho-dynamics, followed by tipping points, eons of drama, transition, extinction, and radical change. We ride the wild Earth that birthed us, but it may well toss us off. Earth from our viewpoint is benign, moody and fickle.

Of all the new paradigms, our dependence upon Earth is the most hated and denied, because many human beings like to take and get more than they like to nurture and revere. We ride on a rock, on a spaceship, yes, but mostly we eat, drink and breathe at the behest of other delicately crafted beings, called "living things." When humankind actually awakens, we will bow down and pray to algae and broccoli. We live on oxygen, and on complex food molecules, that other beings create. There is no such thing as a self-made man. We are totally beholden.

Life seems to have an insistence. We find ourselves surrounded by unprecedented diversity of life forms, niches, and specializations. But this slow parade has also contained extinctions and dead ends: there have been eons and ages without a single face.

XV

Many Forms, One Wave

The great diversity of life, its myriad forms, is the rearrangement of macromolecules due to history, conditions, adaptations, and selections in a four billion year old wave that can raise up atoms and photons into mother-love.

Our sense of wonder is piqued by the panorama of life's charismatic forms. The love of life, "biophilia" as E.O Wilson has dubbed it, in his book by that name, is one of the most frequently cited stimuli to wonder: children and zoos, deer in the woods, zinnias in the garden, us.

Darwin and Wallace provided the breakthrough about how the ark of life became so inventive and diverse. But the spectrum of life's variations remains heavily inexplicable, because which animal adapted to what condition, when, remains a formula full of unknowns. Gould (Chapter Eight) and Eldredge have amplified Darwin's theory of evolution by emphasizing that the historical record reveals punctuated equilibria instead of Darwinian gradualism. (Their theory was first published in 1977 in the journal, *Paleobiology*, and later absorbed into many of Gould's popular books). Life dies off in truncated initiatives, and in mass extinctions, and then explodes in bursts of proliferative invention. There are periods of equilibrium punctuated by wild destruction and creation. The long-term history of life seems more chaotic than careful. Maladaptation and adaptation seesaw, as life first adapts to conditions, but then is presented with new conditions which make the former adaptations archaic.

The world we inhabit is a circus of forms, of oak leaves and kangaroos and human fingers with and without carpal tunnel syndrome. Everywhere we look, a creative thrust has executed a formational tour-de-force. All the diversity in life forms that we can see is only the surface of inventiveness in macromolecular biochemistry that molds the outsides from within. The hidden threads of DNA clothe themselves in three-dimensional expressions of their chemical scripture. The formed world that we see fills out from codes and covalences snuggled in the chromosomes, that hold in chemical language the memory of how our ancestors solved problems.

When we look around us we see millions, tens of millions, or hundred of millions of year-old surging shapes. On September nights we hear the throbbing of tree crickets who also rippled air when misshapen continents were floating and colliding like lost ships. The amazing relationships between wasps and fig trees, in which each of the approximately 750 species of figs is pollinated by a unique species of wasp (there are subtle variations on this) is a relationship that evolved over ninety million years and continues today. We gaze up at a flood tide of molecules interconnected in magisterial and ebullient arrays.

The variety of life is not fully explained either by adaptive gradualism, punctuated equilibria, nor by DNA genetics, and remains a salient stimulant of wonder. Nature, as we encounter it, is the latest iteration of cascades of selections, exterminations, punctuations and proximities. The conditions to which adaptation is a response are themselves elusive, changeable, and nuanced. The laws of chemistry and biology limit and guide what is possible, but a ceaseless flux of conditions evoke a tide of inventive responses: blue whales, elephant trunks, army ants, and John Donne's poetry.

Evolution is too often portrayed as a progression, a sequence. But the bedrock beneath Earth's tectonic plates is constantly moving, the climate is constantly changing, ocean and land have swapped locations repeatedly. The forms around us are only the current phase of billions of years of adaptations to chaotic, gradual, abrupt, climatic, geologic changes. The Earth's variations have no goal or endpoint. In the last several million years alone, ice ages have come and gone as many as twenty times, jungles have dried to deserts, and great herds of giant animals have been born and died in sequences of adaptation, proliferation, variation, migratory radiation, and extinction. Continents have split off or fused, mountains have spiked or shrunk, ocean currents have been redirected by continental channels or blockades. Many funny mammals, many long noses have appeared and exited the stage.

The changing conditions of the planet provoke new responses by which living beings themselves become conditions, which catalyze new adaptations in their neighbors. In his book *Trees*, Colin Tudge has described an acacia tree in Australia as a "network" or "hotel," as animals eat its seeds, bees pollinate it, ants who make it home protect it from browsers with their bites, fungi help their root efficiency, birds spread their seeds, while they send chemical messages among themselves in their groves, and bacteria help nourish them with nitrogen.

Life is like a dialectic. Every species is an answer to questions posed by conditions, and every answer becomes another condition, as life-forms limit, compete and cooperate in colliding existences. Triumph, desperation, and invention temporarily mold molecules into whales.

The creatures of the Earth continue the age-old process of dying and evolving in large numbers. The most drastic extinctions and expansions occurred hundreds of millions of years ago, and involved weird and un-cuddly shells and reptilians, like those of the Permian extinction about 240 million years ago, but recently, even since our own species, Homo sapiens, has existed, there have been significant (though less pervasive) extinctions. Scientists speculate whether the die-off of mega fauna, like mastodons, saber-toothed cats, and giant kangaroos, variously fifty to ten thousand years ago, was due to climate change, human hunting, or other factors. While the trees that form our forests may be tens of millions of years old, the mammals we see around us are often relatively new evolving species, as we are. Biologist Michael Benton wrote in *The Book of Life:* "There are no victors, only provisional survivors." His colleague, Christine Janis, added: There are "hundred of thousands of lost plot lines…and no main highway."

When we think of the green Earth covered with plants, we are actually seeing only one phase, one tidal wave, of Cretaceous angiosperms, the flowering plants invented about one hundred million years ago, which we take for granted, but which are relatively new and temporary inventions.

We love to watch a doe and fawn graze under the apple trees on a summer morning in Vermont, because our eyes are seeing the exact moment when a wave crests, after rolling across time for some four billion years. There is a tide that raised up the doe's mother-love from out of the expanding universe's atoms, sunlight and grass. Doe and fawn are where an unbelievably long and varied history has crested today. Out of the dense interaction of molecules called "deer," love emerges among the leaves.

Every deer, every tree in the forest, is a response, an invention, an expression of a mutating gene thread, and everything we currently see is so old and new, one hundred million years at best. Everything has an ancient pedigree, and everything will be replaced. The forest is breathing and swaying, the most recent formulation of an evolving emanation. It is a wave, a banner, a passing cloud, a long rehearsed anthem, a semaphore of wonder.

XVI

Relative Relativism

We have an expanded set of tools to analyze influence among events and to critique our own ideas, permitting us to maintain confidence in our probing while we remain cautious about our biases.

We have discovered self-reflection, and knowledge about what we call "knowledge." Although some degree of self-critiquing epistemology began with the ancient people of Greece and India, we have in the past century or so become much more attuned to the problem of how we know what we know.

Our knowledge is embedded in the shifting matrix of cultural and historical ideas and instruments, a temporary format that is subject to change. We are wary of any claim to absolute truth. We specialize in toppling what were once seen as facts, realities or verities. We recognize that whatever we say is situated within a vocabulary, grammar, or system of postulates that may be supplanted by subsequent discoveries and formulations.

Compounding this cultural relativism, we have a heightened appreciation for the subtleties of causality. Events may influence other events not only through head-on impact, but by glancing blows, additive or subtractive vectors, pulses of energy, and nuanced probability of encounter and contact. The sciences of cause, chaos, probability, uncertainty, waves, and Einsteinian warps in space and time have alerted us to the numerous ways in which influence between events can occur, and how difficult it is to reach unimpeachable conclusions about relationships between the components of the world. Both the world and knowledge about it meanders. Beyond the readiness to both accept and reject, as discussed in constitutional paradigm VI, above, we have an expanded set of *tools* with which to do so.

At the same time, but with opposite implications, we have improved our skills in *dismissing* untenable belief and false correlation. We can see into dimensions smaller than light waves by harnessing electron motion in electron microscopes, or X-ray crystallography. We launch rockets and satellites into the domiciles of the non-existent gods. We use sections of the electromagnetic spectrum other than visible light to look across trillions of miles of space and backward billions of years into time. We assess probabilities and correlations with analytical tools that require many computerized iterations of statistics. We are more skillful at differentiating coincidence from cause.

Sanctified certainties, forbidden questions, and anecdotal faiths are in the process of being deleted from the curious and reasonable global intellectual community.

We understand that facts can be evaluated by evidentiary confirmation, gathered and analyzed in systematic ways, subjected to public critique, and retested from different samples, different angles of analysis, or varying dimensions and domains. We understand that truths can be supported or challenged by the network of their implications in proximal arenas: narratives mesh or clash within the network of narratives.

We know we prefer antibiotics to treating infections by smearing ashes on our forehead or to sleeping in a scarlet blanket. We recognize repeatability, integrity of observation, bias in selection or report. We are ready to reject our own ideas and to be surprised, embarrassed, renewed. But we are not locked into skepticism or doubts.

The expansion of our testable, repeatable, and interlocking information base during the last century is probably the most revolutionary event since the formation of the first cell. Life—us—can photograph the cosmos from the Hubble telescope and correct some of the effects of errant DNA with designer molecules, like monoclonal antibodies. Human empiricism is a wonder of the self-aware world.

Elinor Ostrom was the first woman to win the Nobel Prize in Economics, in 2009. When she died of pancreatic cancer in 2012, an online obituary by one of her former students, Rick Wilson, a professor at Rice University, claimed that her powerful career of dozens of books, hundreds of articles, more than one hundred PhDs granted, could be summarized by her refrain: How can you prove to me that what you just said is true?

We have replaced assumptions and traditions with investigations. We are not only the tool making, cultural animal, but also the persistently probing one. And very rarely, someone recognizes the belief that "all narratives are culturally relative" is a culturally relative narrative.

Somewhere there may be a wall, a threshold, a gold key, a group of Gödel-impervious axioms, an ultimate particle, a unified field theory, a guiding hand.

Every sane person notices that their mind and their heart, though related, are not exactly the same thing. Like any good couple, they neither stifle nor submit to their partner, but tease and play as long as their life endures. We think and feel and do not know, and yet we still try to know, oscillating between intuitions of the ephemeral and the eternal. Ignoring facts is the route to self-deceptions. Adherence to facts alone will stifle the most fertile zone of hypotheses, the heart.

XVII

Living Cosmos

The universe underpins and permits life,
of which we are a local manifestation.

The cosmos has an aliveness. The qualities that we associate with our own lives are properties of the universe that we have downloaded, condensed, and intensified. We are partly specific to the weaving of laws and events on the platform of Earth's four billion year history, true, but nothing has happened here that is outlawed or impossible. The universe not only permits, but in zones and domains augments the qualities that underlie our life: lawfulness, limits, channels, and constraints; plus change, variation, uncertainty, indeterminacy; plus interconnections, interactions, influences, and communication.

Connections and relationships in multiple domains and in woven complexity, at high speed and in huge numbers, over long eons, raise up out of matter, energy and information the potential of life.

The origins of life on Earth is an intriguing scientific mystery well worth studying, but it should not be mistaken for the origin of life. Knowledge about brain structures that are essential for the manifestation of consciousness may help us deal with important psychological or medical issues, but they are not the origins of consciousness. Life and consciousness are not exemptions from the cosmos but expressions of it. The human mind did not create consciousness but rediscovered how to raise it up.

Our enriched sense of reality has brought to our attention the limits of machinery as an image for reality. The reductionistic sciences, that model the world as interlocking parts in linear connection, has been helpful in dispelling superstition, but does not help us explain many phenomena from quantum mechanics to creative evolution. When we study life, we find a world never at rest, cascading and surging in innumerable interconnected infinitesimal relationships. The relentless penetration of thing upon thing in the micro-spaces of cells, the resonances from the cosmic corona where energy and matter merge, remind us that life touches down into the essential features of the cosmos. The sciences of complexity, indeterminacy, and of life itself make us feel most scientific when we are whirling in mixed influences and slightly off balance (or, as it is called in science, capitalizing upon broken symmetry).

We are neither a capricious violation of cosmic law, nor a pre-programmed, designed creation. We are not wedded to grim and alienated atheism, nor to infantile father projections. We do not expect to encounter faces like ours waving to us from black-hole drowning galaxies, (because our forms are not only based on cosmic law, but also upon Earth-evolution's unique circumstances) nor do we accept being assigned to the stature of accident.

Our internal galaxies of atoms are writing one volume in the library of the universe. What the kangaroo and koala are to Earth, we are to the galaxies. As ecosystems are to plants, the cosmos is to Earth. We are context-dependent in a Chinese-box set of contexts. We are a lot of properties and potentials encircled in one place and time, impressive and slightly funny. As much as space, time, matter or energy, another property of the universe is life. The secrets of the universe are not different from us. Our specific form is a local accident, but our essence is essential. When enough waves converge, a standing wave forms.

XVIII

Human Evolution

*We are new, evolving, and in the process of cultural transformation
due to our awareness of the galaxy and the cell.*

At last we come down upon the evolution of us. When was the first human? To some extent the answer is definitional. Upright apes, hominids, were around five or six million years ago, but significant tool use had to wait for our own genus, Homo, to appear, several million years ago. Our species within the genus is much newer, and the specific group of Homo sapiens who are the purported ancestors of most of us were present less than two hundred thousand years ago, (often said to be one hundred thousand), lending their genes to us, as we dispersed out of Africa about fifty-thousand years ago. (The theory of Homo sapien evolution continues to change. We may now be considered partly the dominant,

African-derived species, and partly descended from European Neanderthals and Asian Denisovans). You could say we are five million, or two million, or two hundred thousand, or fifty thousand years old or new, at the tail end of the approximately four billion years of evolution of life on Earth. It takes a long time to craft a good ape, or a representational mind.

Our common humanity can be felt in the way we touch the world. Anthropologist Loren Eisley wrote in *The Firmament of Time*, that when he peered into ancient gravesites he felt a kinship with even million year old bones because of the human gestures of loving burial, art and tools. Even these pre-Homo sapiens "...had laid down their dead in grief." Poetry from other eras still speaks to us. John Donne's "Anniversary," with its archaic spelling and grammar still claims its sacred hold on the triumph of romantic love:

> All other things, to their destruction draw,
> Only our love hath no decay...
> But truly keeps his first, last, everlasting day.

Though we can feel our continuity with previous ages of humanity, we also have large differences. We are evolving culturally as well as biologically. No one else before us has had a clue about what the stars are. No one else but us has had an inkling about the chemistry of cells. Earlier Homo sapiens had tool use, but no orbiting space telescopes or electron microscopes! John Donne still speaks for our loves and hopes, but not for our context inside a brand new evolutionary realization. We have broken through fourteen billion years of opacity. We send out faint oscillations, of which the chemistry of our thoughts consists, towards distant galaxies. Earth life, for the first time, can send thoughts towards and receive signals from other galaxies.

The most likely planet that we have seen to possibly harbor life is a mere twenty light years away, a mere one hundred and twenty trillion miles or so, although the universe is more than ten billion light years big (sixty billion trillion miles, give or take, or maybe four or five times bigger, as mentioned in Chapter Fifteen). We can for the first time conceive that the universe contains more dimensions and possibilities than we can hold easily in our minds. We are the only Earthlings ever to understand the multiplicity of potential life platforms within which we float. Understanding and tinkering with our own DNA creates a chasm between our ancestors and us, for we can dream of participating directly in our evolution.

It is not unusual for a species to come into being (or to die off) in the course of a million years: maybe we have only just been reborn via cultural evolution. Our discovery of the galaxy and the cell frame our mind. We are a new being. Have we become a new species? We have the old emotions, the spear points and the sorrows, but we have entirely new thoughts. Our representational mind—that can envision, imagine, simultaneously hold up alternative pictures, think before action, and construct narratives of the future in a decision about how to act now—that mind can now represent radically new intangible micro and macro worlds.

Our new thoughts in turn derive from and propel new configurations of spatial brain chemistry. As we probe the galaxies and the cell, our new awareness may reconfigure our subtle chemical architecture. Our brains may be modified by our insights. Our wonder may set an evolutionary trajectory as catalytic as hunting and farming once did. Our knowledge about the arrangements of distant lights and nucleic acid sequences may change us more than we can now know.

Consider the three to four billion year journey from bacteria to human. Now consider a four billion year journey beyond the human, with concomitant increase in complexity of DNA, multicellularity and neuro-circuitry. What will that being with trillions-fold increase in cognitive capacity beyond ours understand about the information base of the universe, its origins, meanings and goals? Then consider the evolution of the society and culture of these neo-knowers, and what their libraries and Internets may contain and understand.

We try to understand our world with a half-way-there body and mind.

Are there in fact beings on a much older planet, that had much more time for evolution, rotating around a much older star, who manifest many more potentials and properties of the universe than we do, including insight into who they are?

XIX

Population Dynamics

*Human beings appear to be in a process of natural, biological
self-elimination through resource depletion and overpopulation,
which rationality and social feelings may or may not prevent,
unless we become both more realistic and reverent.*

Basic biology teaches us that if amoebas are inserted into a drop of water, their numbers will expand, until they become so densely populated that they deplete their essential nutrients, and die *en masse*. The drop of water again becomes uninhabited and sterile.

We humans are doing the same thing on planet Earth. We use up too many resources with our automobiles, meat-eating, and petrol-industrial agriculture; we are too numerous with our seven billion people who are expected to reach nine billion within the next lifespan; and we are too irrational, creating wars about gods that do not exist, and letting perverted old men block the universal option of contraception and family choice. We are eating every last fish, and sequestering a nation's worth of resources in one man's private bank account.

We have yet to prove that we can escape from the population dynamics of amoebas. Malthus warned us in the nineteenth century; Paul Ehrlich reminded us of the population bomb; both were dismissed as cranks, and neither has been disproved. We have yet to harness cause and effect thinking and probable outcomes into our collective planning. We have yet to design governance that modulates power, greed, delusion, fear and xenophobia, into widespread and shared well being.

Yet the sun showers us with six thousand times more energy than we use. Solar energy could supply a human population of trillions. The insecurity and loneliness that drive status seeking, power, and hoarding can easily be transmuted into participation and caretaking, for most people have in fact already attained this. Most of the seven billion of us are basically social. Both terrorists and eco-destroyers are rare though devastating contagions. Today we know that we are all an African family on a long camping trip.

As a general rule, species last only so long. Most species last around ten million years, and most disappear without leaving any derivative species. Most are blind alleys of evolution, roads not taken. The history of life predicts our demise as a species. Species die out when niches are eliminated due to climatic or geologic change, or when out-competed by other species (which seems unlikely for us), or when overpopulation leads to niche depletion and extermination, as seems to be happening. In the fifty thousand years since the ancestors of most of us left Africa, we have exploded in a manner that predicts contraction.

It seems hard to accept that we are only like the millions of species of barnacles and snails that died in spectacular extinction events recorded in the fossil record, or who disappeared in singular, species-specific extinctions, like the elms of England. Our explosion of knowledge has led us to consider how limited our time span may be. The diplodocus may have thought that he was here to stay. Why is it so hard for us to agree to drive efficient cars, eat less or no meat, limit our families to two children, and grow our corn and soy beans locally and sustainably? There may be many kinds of deep-sea marine invertebrates and underground cicadas who are guffawing.

The food grains that feed us (or that feed our chickens and cattle), like corn, rice, and wheat, are derivatives of grass, a relatively new plant, only some thirty to fifty millions of years old.

In order to prolong our presence, we need to cultivate and revere our real saviors. We need to recognize our wide-ranging kinship amidst impermanence and change. We are still children of a rainy day that greens the grass. Everything we eat is a product of sunlight and photosynthesis. Every raindrop is worthy of worship.

From our moments of wonder, let there be a harvest of realism and reverence.

XX

Projection and Awakening

Human beings project the emotions of their past onto the present,
distorting perceptions, which could also be
cleansed into freshness and wonder.

We identify with our emotions but they can be misleading. Twentieth century psychology has taught us to both cherish our emotions, while we also recognize our tendency for projective distortion.

Our emotions typically feel more intimately embedded in our sense of self than our ideas do. We feel we *are* our feelings. All reasonable people recognize that their ideas are subject to error, and that our choices need to be guided to some extent by feelings. Woe to the man or woman who marries based on an idea! Elizabeth Kubler-Ross, in her book, *On Death and Dying*, demonstrated that peoples' *ideas* about faith, belief, or un-belief had little correlation with their real *feelings* at death. Their ideas and beliefs were only claims; their feelings had more weight as they were dying. Our emotional intelligence may predict better than our I.Q. our ability to construct a high quality life. We feel we are true to ourselves when we make heartfelt decisions and not merely practical ones.

But our feelings can be misleading, being partly those of a mammal, a primate, whose close relationships control much of what we feel. People may feel rich or poor not because of what they actually possess, but in comparison to their neighbors. Our local pack status, our location in the hierarchy of power and dominance at home or work, may dictate our stress hormone levels, or how we feel about ourselves, more than "objective conditions" do. Our early childhood experiences with parents or caregivers may fix us in a band of relatively good or bad mood for life. So may our supplies of certain brain chemicals. So may one-time, indelible traumas that entrain our physiology and chemistry to states of terror.

We *feel* the world is an extension of our previous personal relationships. We project onto situations, the universe, gods, the feelings to which we have already become accustomed. Feelings are easily habituated and generalized. We feel the world is provoking us to feel as we have already felt. We feel about life based on a faulty sampling procedure, assuming that our limited past experiences predict the future. As parents stood over us and guided us for the early decades of our consciousness, so we imagine the black and shining cosmos does. Because we powerfully experience our personal desires, we imagine the world caters to and coddles us. We imagine that the intelligent beings, long nurtured on ancient planets rotating around ancient suns in galaxies that matured before ours was born, are cuddly stuffed toys who have electronic voices and drive hovercrafts, similar to the toys that populated our shelves when we were growing up.

Freud coined the concept of "projection" to describe situations in which a person denies his or her own unacceptable urges, and attributes them to someone else. We have come to think of projection as a more prevalent phenomenon of smearing onto our image of the world many of our past experiences. In his seminal dissertation about religion, *The Varieties of Religious Experience*, William James described people as having a total reaction upon life that appeals to the person's temperament, and is therefore preemptively biased. The Buddha took a similar position in the *Mahasatipatthana Sutta* and elsewhere when he described meditation as (among other things) the cessation of conditioned perceptions and judgments based on the past, and a simultaneous direct experience of subtler and more immediate realities. Social psychology has demonstrated how group expectation and culture impact our perceptions of things that we would like to believe we know deeply by ourselves, even right down to our personal assessment of our own happiness. Our feelings and perceptions are tricky genies.

Projections are not always accompanied by words or thoughts. Our emotional life may revolve around moods that were developed in the past and that we now feel willy-nilly during new experiences and circumstances. Both religious and secular scientists may project onto the world their emotional states, even in direct contradiction to their knowledge. They may feel there is a father in the space whose real galactic eternities they know; or they may feel a deadness in a world that in fact proliferates toucans, John Donne's everlasting day, and Albert Einstein's wonder.

If the projections of childhood, familial, self-absorbed feelings are released, we might awaken. Stunning magnitudes and inheritances surround us in the scintillating world. Then we may well sing and worship. When we notice and relinquish our projections, many modes of knowing and not-knowing may open for us. We will be dumbstruck by marigolds.

SECTION VII

Light and Dark Wonder

Chapter Seventy-Eight

Light Wonder

A t last we come around to the compounded world of daily events and textures. How can we retain our vision of wonder in the domain that lies between molecules and stars? Let's start with ice.

I am looking at a grillwork of icicles hanging like daggers in front of my office windows on a cold winter morning. They sparkle so blindingly that, like Whitman, I only gaze at them from time to time (Chapters Two and Twenty).

Each icicle answers to the laws of physics by which, at temperatures below 32°F, water molecules abandon their aqueous phase-state and flip into a new self-organizing pattern. The icicles also obey gravity, which draws the alternately freezing and sun-thawed drops down towards the center of the Earth, so that a curtain of sparkling stilettos festoons my windows. These icicles are basking in glorious electromagnetic radiation sent to us by the sun from more than ninety million miles away. As the photons collide with ice, rouge diamonds refract across the curtain of cutlasses. Although this blazing gallery is a product of the constitution inside of all things, the local interpretation of the codes is lax. Each icicle is weird, jagged, unique. Each weapon is being remodeled in front of my eyes as the cold air and the photonic electromagnetism spar by freezing and thawing, while gravity like an umpire grounds each drop. Subtleties of location and angle proliferate variation in this cosmic cutlery of ice and light. How can an impersonal universe create such radiant beauty?

When I look away from time to time, I notice two red-tailed hawks perched on the upper limb of a poplar on my neighbor's property. The hawks sit like married statues on the battered, high limbs, where they are probably warming themselves in the photonic waves that turn to heat upon collision with the bird's white abdominal feathers. Mr. and Ms. Hawk must also be alert for prey, which can only be acquired at a high energy price in the snow-bound landscape.

Everything in this tableau of apparent stillness is actually in motion. The icicles morph as I watch them, occasionally plummeting down into the snow. The big birds are not still but poised, and they are set to lift off in pursuit of their caloric necessity. My mind too is rippling in wonder. Man, bird, and ice all contain in their water, hydrogen atoms from the dawn of time. The red-tails carry within them similar information in the same genetic code as the voles they eat, a code assembled three to four billion years ago and first reflected in the human mind a mere fifty years ago in our cosmic yesterday. It was not until the 1960's that teams of bio-scientists, including Francis Crick, Marshall Nirenberg and many others, first mirrored in their consciousness the informatic sequences that form DNA codons, so that life at last became aware of itself as partly sequential information.

Man, bird and ice are cosmic conglomerates, temporary residents, artworks in the same installation. In this small crucible of space, in this coincidental birthday party of making and melting, a universal artist plays.

I wonder whether there could possibly be envy and competition among cosmic artisans. Could planet Earth flaunt its magnificence of red-tails to poor, hot planets blazing fecklessly in spiral galaxies devoid of Francis Crick, Marshall Nirenberg, Walt Whitman and red-tailed hawks? Is the mind inside a hawk similar to my mind, a variation upon it, or a kindred and essentially similar Morowitz-and-Pauli-sort of universal mind (Chapter Five)?

My speculations halt as the hawks fly off, the orchestration shifts, my own great ship of Earth sails on through elastic-black space-time, changing its angle to the sun. The icicles suddenly look dull gray.

Inside the polymath potential of the universe are the capacities for men, birds and ice. The creation never stops nor is still. Flows form and unform. Law, history, change, and circumstances guide them. So many things have been created and in one golden moment I was struck by the wonder of it.

In order for this moment of wonder to have occurred, here I sit in my office, as my planet rushes through space at about five hundred thousand miles per hour (there are many motions in this approximation: the Earth revolving around the Sun, the Sun rotating around the Milky Way Galaxy once every two hundred and twenty million years, the galaxy sweeping forward with expanding space), through the caverns that are (at least) fourteen billion times six trillion light years wide, my planet having given birth to life some four thousand million (billion) years ago, and having created in that long interval about fifty billion species (see Chpater Seventy-Four), and having covered itself with oxygenated plant-breath, and having ushered in oxygen breathers, and having elaborated in the play of life over the most recent hundreds of millions of years complicated mammals, who have increasingly complex minds, all of this swirling aloft in my one hundred trillion cells, one hundred trillion informatic DNA threads, septillions of ATP's and thousands of other smart molecules, containing octillions of atoms whirring with coordination and precision and whizzing accurately among themselves across uncountable nano-gaps, while creation and destruction bring life and death to me with their roughhouse game.

One moment of icicles and birds pours out of the cosmic fountain. One moment of awareness of icicles and birds is more, and different, than all of its parts. There is no resting place and the scene moves on.

Chapter Seventy-Nine

Calling of the Heart

The season and the scene have changed, and today a woman is planting flowers in her spring garden. Her activity is more than a hobby, even more than a pleasure. It has the intensity and absorption of childbirth. It is for her a necessity, an outcry, a response like biological reproduction. She is digging, dirtying, straining, mulching and lugging, under the power of plants which do not yet even exist, but whose images have taken up residence in the atoms and cells within her imagination. Their seeds have penetrated her thoughts and sewn them with visions of petals. In her spading, mixing, and tamping down, flowers are coming to be. Though she experiences herself as a self-directed person, in fact, flowers have possessed her, and she has become their vehicle. She is a woman in flower birth.

Weeks or months will elapse before her labor is fulfilled. Patience and faith will sustain her until, under the majesty of Earth's dominion, the unprepossessing little bulbs and seeds will explode into daffodils, tulips, irises, bee balms, delphiniums, geraniums, pansies, daisies and sunflowers. Their diaphanous hues will tint a small corner of the black entropic universe that must now make way for a summer dress of petals.

Jacques Monod, the Nobel Prize winning scientist and French existentialist (Chapter One) dared to stand firm and tall, alone in the unfeeling immensity of the universe, but now that same universe has parted its midnight curtains to reveal an azure and ruby décolleté.

A woman, flower seeds, rain, sunlight and wormy soil have beaten back the mighty vultures of emptiness and entropy. A war has been won by soft and colored things. The yellow eyes of asters, the purple tongues of irises, and the crayola pansies have raised their banners above the turrets of Earth's soil to defy the dark cold space that pervades almost all of everything else.

If there were a heaven, the gods would abandon it just for the chance to see this woman in her garden. The great old theme of light and dark has its denouement here. Her jonquils have feasted on sunbeams and claimed the field. Here the universe has rolled and kneaded all of its complexity into pastel translucence. In the mixer are electromagnetism, electron rearrangements, chemical bonds, DNA sequences, Earth's plate tectonic climate, photons, Cretaceous flowers, mammalian evolution, and human culture. Hoary eons converge upon the color, yellow. The rough and ancient edges of time have rasped out an unexpected communion between human and plant, woman and sun, mother and mulch, girl and petal. Although flowers originally served as adaptations to attract bees, flies, birds, bats and other functional pollinators, they also attracted this woman, who, like a hummingbird, is drawn to their sweet centers of color and light.

Is she growing flowers, or are the flowers just using her as their grounds' staff? Sunlight has succeeded in pulling out of its hot vaults this symbiosis of peony and woman. How have all the eons and forces *aligned* here at this moment to grow daisies with their gold pendants and white lace? Is it possible that flowers are as foundational as the sunlight that their leaves lick for life? Has this woman chosen gardening, or was her lengthy and tailored evolution an assignment? If the late Steven Jay Gould were to have imagined this garden as a quirky happenstance that could just as easily have never happened, would he continue to believe so, if he saw this one hundred pound wife drag a fifty-pound sack of mulch towards the tulip bed like a soldier rescuing a wounded comrade under fire?

There is a force, a combinatorial imperative, that begins with quarks and their kin and that does not halt after molecules have been assembled. Quite possibly we love by the same spirit of combinatorialism that is so greatly esteemed by quarks, carbon atoms, and endosymbionts. As humanity has become aware of this force, we have labeled it "metta" in Buddhism, "agape" in Christianity, "love" in common speech. It is not merely a survival-oriented adaptation. It is not merely reproductional. It may well be that a woman in love with flowers is a manifestation of as fundamental a law as is electromagnetism or quanta. Our gardener has risen up from billions of years of woven laws like Botticelli's birth of Venus from the ocean. It was sunlight that warmed our planet and supervised the loom that wove our gardener and all her flowers during Earth's four billion years. It is sunlight transformed that powers her flowerstruck mind.

At this place and moment, this woman and her garden are the last stop for the energy of hydrogen fusion, the terminus of the Big Bang's big train. All the laws of the universe had to be involved in order for her to abrade the knees of her jeans with silica and mica of the New England soil. When she carries bulbs from place to place, her small muscular frame casts shadows on the rotating Earth like a galactic Amazon, a time traveler striding from the Big Bang to the Milky Way. If there were an initial information state at the dawn of time, it has been shredded, re-sewn, quilted into a message of marigolds. The universe has made beings who cultivate beings. Interspecies birthing is a sequence at the high tide of the world. It takes the history of the universe to grow a gardener.

Things appear to us to combine because they are in fact united by long sequences and relationships. It appears to us that a woman (separate being) is (choosing) to plant a flower (separate thing), when in fact, seen from the vantage of space and time, the age old saga of life-forms—ferns, bacteria, dinosaurs, or apes—surges over the surface of the planet in sequence, interconnection, creation, destruction, variation, selection, transformation. What appears to our own parochial perceptions to be discrete entities and unique events are actually embedded in ongoing flow. Matter, energy, information, and dispersion form a sequence as smooth as my hand as I write this page.

That is why our gardener will not stop for lunch. Spring has no synapse. The universe flows on down hill with a force like gravity, with a force like love, with a force like combinatorialism and entropy. Analogous to the way that electrons fall down the radiant scaffold of energy release during respiration, (Chapters Fifty-Two and Fifty-Three) the marriage of springtime and humankind falls downhill into patches of delphiniums and geraniums.

The love of flowers was shot out of the Big Bang in invisible and disorganized array, and it fell for fourteen billion years to land cat-like on its feet in my own backyard. The Big Bang has screeched up to its train terminal in the form of a woman weeding daffodils and soaked in love-chemicals in her mind. And it is hydrogen fusion that provides the energy for the mind of the man who watches and worships the gardener.

The snarly old physicist embedded inside the "initial information state" has gradually evolved into millions of locations on planet Earth where people are listening to the calling of their heart.

Chapter Eighty

Dark Wonder

"...the great flood gates of the wonder-world swung open..."

That is how Ishmael, the fictional narrator of *Moby Dick*, described his voyage on the whaling ship, Pequod. In *Moby Dick*, Herman Melville created the great novel of wonder. The book repeats the words, "wonder" or "wondrous" like a refrain. We are told about wondrous depths of the ocean, the wondrous old man Captain Ahab, the terrors and wonders of God, the wondrous period "before time" when whales left their fossil skeletons "along the present line of the Andes and the Himmalehas (sic)," the wondrous whale itself; the most wondrous phenomenon which the seas have revealed, the squid; the wondrous power of the fin whale, the wondrous devices that whalemen dart over wonderful distance to kill whales, wonder-freighted waves, wondrous philosophies, whales' wondrous breaching, and natural wonders. As if someone had admonished the author not to overuse a single term, we are also recurrently regaled with "miracles," "marvels." and "magnificence." In *Moby Dick,* wonder ceases to be a word and is used as an incantation.

Melville's style is also wondrous, with its manic and mellifluous verbiage, wild and unruly associations, globe-spanning references, classical learning, and his bending and squeezing English usages to elicit our gasps of disbelief.

But *Moby Dick* is esteemed as the great American novel due to the story of Captain Ahab, who lost his leg in a previous attempt to kill the legendary white whale. Ahab pursues Moby Dick with obsessed hatred, even though readers are warned from the start that imprudent whaleboats could be "stoved." The plot tension does not spring from surprise. Ahab's character has been hailed by literary critics as a keen premonition of Hitler. No reader doubts what will happen to mad Ahab and to anyone who obeys him.

How does a book about wonder also contain a story about suicidal-homicidal leadership? What is the relevance of *Moby Dick* and its violence to our contemporary, wonder-filled journey through the universe?

Melville's wonder-world weaves together the story of whaling—Captain Ahab, his crew, and their violent showdown with the great white whale—with the story of Ishmael/Melville and his search through "shore-lessness" to find revelation. Melville worked out his own anguish in this novel. As his only literary friend, Nathaniel Hawthorne, said (as quoted in Andrew Delbanco's *Melville: His World and Work*): "[Melville] can neither believe, nor be comfortable in his unbelief, and he is too honest and courageous not to try to do one or the other…he has a very high and noble nature…." The fictional narrator, Ishmael, became Melville's own voice.

Moby Dick is important to us because it has guts. Wonder is not for weaklings.

Melville evokes wonder as *salvation* for the "unshored," modern mind by demonstrating that wonder encompasses violence, cruelty, destruction, annihilation: *dark wonder*. It was wonder that steered him into darkness in 1851, and inspired Melville to write a novel that was relevant for an era that was exterminating whales and buffalo, chaining and whipping millions of black Americans, committing genocide against Native Americans, and on the verge of "Mr. Lincoln's war" that would kill half a million American men, while "tuns and tuns of leviathan gore" was used "to illuminate the solemn churches that preach unconditional inoffensiveness…" Like the *Bhagavad Gita* (Chapter Two), *Moby Dick* bequeathed to us the capacity to integrate our sense of wonder into a sometimes horrible world. Here is a sense of wonder that is enriched and not deflated by grim realism. "That mortal man who hath more of joy than sorrow in him, that mortal man cannot be true…"

After Captain Ahab, the Pequod, Ishmael and the reader have pursued Moby Dick for five hundred and forty-nine pages, they arrive at the vortex of wonder and violence:

"A gentle joyousness—a mighty mildness of repose in swiftness… the glistening white shadow…not that great majesty Supreme! did surpass the glorified white whale as he divinely swam…through the serene tranquilities of the tropical sea, among waves whose hand-clappings were suspended by exceeding rapture…withholding from sight the full terror… the wrenched hideousness of his jaw…the grand god revealed himself…"

The god-like whale easily evaded his would-be killers, but instead of disappearing, in an act of "defiance," he "booms his entire bulk into the pure element of air, piling up a mountain of dazzling foam" in his "wondrous phenomenon of breaching." Serenity, majesty, and wrenched hideousness combined in the great white whale's portentous wonder.

Melville loved and revered real, non-allegorical whales too, and tried to explain their "wondrousness" in many descriptions: "appalling beauty… impossibly beautiful…exceeding grace…maidenly gentleness…then smiting the surface, the thunderous concussion rebounds for miles…this peaking of the whale's flukes is perhaps the grandest sight to be seen in all animated nature." Watching a pod of whales heading towards the rising sun, he proclaims them as "a grand embodiment of adoration…" and "… the most devout of all beings." He ends this description with his trademark irony: "The mightiest elephant is but a terrier to Leviathan." His description of female whales with their calves erased any distance between animal and human, for in the "submarine nurseries" he saw "…as human infants…still spiritually feasting upon some unearthly reminiscence… baby's ears newly arrived from foreign parts…" He tells us of "The inherent dignity and sublimity of the sperm whale…Heaven itself put its seal on its thoughts." Even murderous Ahab revered whales and saw them as "… homage-rendering, faithful, broad, baronial vassals of the sun."

This love of whales only played into Melville's wish to horrify us with their slaughter. *Moby Dick* contains repeated, pounding descriptions of the cruelty of whaling. If we want real wonder, honest, enduring wonder, Melville seemed to say, then we must hold it in the context of whaling gore.

As Ishmael/Melville "lay entranced" watching nursing whales, the "war" commenced, with the attitude: "…you must kill all you can. And if you cannot kill them all at once, you must wing them, so that they can be afterwards killed at your leisure." Mothers and children aren't spared: "When by chance these precious parts in a nursing whale are cut by the hunter's lance, the mother's pouring milk and blood rivaling discolor the sea for rods." Lest we accepted the massacre as happening only to big, dumb beasts, Melville told us about "…the extraordinary agony of the wound… an appalling spectacle…peculiar horror…tormented to madness…the red tide now poured from all sides…His tormented body rolled not in brine but in blood which bubbled and seethed for furlongs…jet after jet of white smoke agonizing shot from the spiracle of the whale…with sharp, cracking, agonized respiration." We are told how the whaler "…churned his long sharp lance...and kept it there, carefully churning and churning...he sought the innermost life..." until the whale went into "...that unspeakable thing called his...flurry...horribly wallowed in blood and impenetrable, mad boiling spray..." Melville was not content to upset us once, but kept repeating similar scenes "unspeakably pitiable...the utmost monster of the seas was writhing and wrenching in agony...showers of gore." Melville stated: "...there is no folly of the beasts of the Earth which is not infinitely outdone by the madness of men." The "deeper midnight" that whaling revealed, "...would make an infidel out of Abraham."

Amazingly enough, the same author was able to convince us of the remarkable courage, ingenuity, hardiness, and globe-conquering significance of whale-men. Melville stated that he, and we, all are killers, and yet he wanted us to behold our own species, too, in wonder. To do this, he reversed the power of his anthropomorphisms. Whales were elevated above people, and people were degraded to the level of sharks—whalemen on the deck feasted on whale-meat, and sharks feasted on the floating carcass, and if "...you were to turn the whole affair upside down, it would still be pretty much the same thing..." This degradation of humans to the level of beasts also allowed Melville to wonder at the inventiveness and creativity of killers. He dared to see humans as refined predators, exemplifying natural law. After all, compared to the unspeakable carrion of the battlefield "...whale ships are clean, constructive sources of the lamps that made nineteenth century civilization possible...almost all the tapers, lamps, and candles that burn round the globe..." "The high and mighty business of whaling...has been the pioneer in ferreting out the remotest and least known parts of the Earth...the mysterious, divine Pacific zones the world's whole bulk about..." The Pacific-bound whale-ships stored in their hull "enough water to form a lake," enough to last the three or four years journey. The American whaling fleet was created by the tiny population of Nantucket Island, whose heroic men, "...these sea hermits issuing from their ant-hill in the sea, overrun and conquered the watery world like so many Alexanders...let the English overswarm all India, two thirds of this terraqueous globe are the Nantucketer's." They had a business that "Noah's flood would not interrupt."

The Pequod was also manned by a racial diversity that rewarded excellence, unlike the racist, slave-owning, suicidal-homicidal, war-bound U.S.A. of the 1850's. "Not one in two" of the crew were American born, and all three of the fictional Pequod's harpooneers (sic) were non-whites, yet "it is the harpooneer that makes the voyage." The famous Queequeg (somewhat incongruously described as a South Sea Island Prince who seems to be part Moslem, part Buddhist, and part "cannibal") is "…a wondrous work in one volume…with an innate sense of delicacy and so much civility." As America tortured its African Americans, the whale ships mixed the races and depended upon all.

Melville described in meticulous detail the rendering of a dead whale into oil that would light New York and Boston, and was dazzled by the inventiveness, efficiency, and laborious ardor involved. His descriptions leave the reader wondering how such inventions as the "try works," that boiled blubber in great vats on the deck of a wooden ship, could ever have been invented. Whaling was "a vocation of perils and wonder."

But Melville's greatest sense of wonder was saved for courage. As the whalers departed in thin, small boats from their mother ship, in order to approach whales close enough to throw hand held harpoons into them, and then pursue in the open ocean dangerous and wounded Leviathans, Melville's admiration and wonder expanded: "…here goes for a cool, collected dive at death and destruction." "Six men composing the crew pull into the jaws of death, with a halter around every neck." "It was a sight full of quick wonder and awe!" Nothing can feel "…stranger and stronger emotions than that man does, who for the first time finds himself pulling into the charmed, churned circle of the hunted sperm whale."

Starbuck, the chief mate had "…dried up all of his physical superfluousness…" prepared to endure "…a thousand-fold perils calmly confronted…" with "original ruggedness…" Starbuck would drive his fragile open boat into the squall, mayhem and "ringed lightnings" of a thrashing giant Leviathan.

"Courage was one of the great staple outfits of the ship, like her beef and her bread…The courage of this Starbuck…must indeed have been extreme…man, in the ideal, is so noble and so sparkling, such a grand and glowing creature …that immaculate manliness we feel within ourselves."

The physical courage of Starbuck was not the same as psychological courage to face truth. *Moby Dick* became focused on the wonder that consciousness and awareness can exist in a world of murderous men. Human virtue and wisdom coexisted within the "heartless immensity" of the oceanic world. While nature pursued creation, it also pursued destruction—Ahab hunted Moby Dick and destroyed his ship, crew, and his own life—

and sharks infested the sea. But Ishmael, Melville's fictional embodiment, who constantly mocked and rejected religion, stared into it all, absorbing, wondering.

Starbuck, in contrast to Ishmael, passively followed Ahab, even though he knew better. Starbuck had physical but not the required emotional courage that we need to find wonder and salvation within a shoreless, cruel world.

Moby Dick was ultimately an invocation of Melville's dark wonder. In a world of demons and dictators, slavery and genocide, in a shoreless and unfathomable world, Ishmael and Melville stayed afloat upon psychological courage. It was the ocean, "landlessness," "the problem of the universe" that created the most terror in humans and that required the most daring for us to confront. While lesser characters confined themselves to their pipes and harpoons, Ahab, who was not a mere cardboard bad-guy, but had "archangelic grandeur," and was "…ungodly, godlike and used to deeper wonders," hated particularly "the inscrutable thing." Ahab's hate was for the world, its gods of justice and injustice, its wonder, its indeterminacy.

In contrast, fictional Ishmael, and Melville, did not hate the unknowable. They contained the courage to float on it.

Using Ishmael's voice, Melville wrote a modernist Bible: "All deep and earnest thinking is but the intrepid effort of the soul to keep the open independence of her sea…in landlessness alone resides the highest truth, shoreless, indefinite as God—so, better it is to perish in that howling infinite…" than to surrender to mere comfort or ignorance. He learned "…a bold…and lofty language." From "awful lonesomeness…the heartless immensity…when the ringed horizon expands…" that carried his thought down into "*wondrous* depths…unwarped primal world…So Man's insanity is heaven's sense; and *wondering*…man comes at last to that celestial thought, which, to reason is absurd…he feels uncompromised, indifferent as his God." (Italics added)

The ocean revealed to Ishmael heartlessness in creation, drove him to altered states of consciousness beyond belief or disbelief, and left him with God-like equanimity. "Amidst the tornadoed Atlantic of my being…I still disport in mute calm…deep inland there I still bathe me in eternal mildness of joy."

Despite his revelations, Ishmael reminded us that wonder required an accompanying alertness and competence. With irony Melville wrote about the experience of surveying the empty ocean from the look-out perch at the top of the mast-head: "To a dreamy meditative man it is delightful…there you stand, lost in the infinite series of the sea…the tranced ship rolls indolently on…sublime uneventfulness invests you…serenity of those seductive seas." But the character Ishmael, echoing Nathaniel Hawthorne's lines about Melville, has "…the problem of the universe revolving in me…" and so he warned the reader not to be "…lulled into such an opium-like listlessness of vacant unconscious reverie…blending cadence of waves with thoughts…" until at last the lookout might lose his identity and mistake the ocean for his soul, and "In this enchanted mood…become diffused through time and space." But wait! You are rocking high above a ship, clinging to a masthead! "…slip your hold at all and your identity comes back in horror. Over Descartian vortices you hover…Heed it well, ye Pantheists!"

We may have deep cosmic identity, but we have bodily physics as well. If your philosophy claims you are "One with Everything," try a shark infested tropical sea.

So it is that we thrive on a mix of Dylan Thomas's cozy wonder, Einstein's scientific wonder, Chandrasekhar's mathematical wonder, Carl Sagan's quasar and black-holed wonder, Whitman's bumptious wonder, and Melville's "ungraspable phantom of life."

Moby Dick addressed fundamental and omnipresent principles. The destruction of entire galaxies inside quasars may in some ways resemble the final victory of the great white whale. The remodeling of cells through constant breakdown of parts may look to the component atoms like assaults by tiny intracellular white whales. So our own destiny may appear to us, on the day that our own nemesis breaches in front of us.

Our wonder in life is often muted by our difficulty accepting reality as it is. Dark wonder is the ability to accept destruction as the counter player to creation. Brutality, mass murder, and heartless gore can break our spirits, and yet some, like Melville, witness it and shine on.

The dynamics of destruction, followed by renewed creation, without rest, in all sizes and domains, is the source of wonder.

References

- ✦ References are designed for an internet era text.
- ✦ All quotes and direct references are noted by author and title within the text itself.
- ✦ Scientific topics, such as protein conformation, chaperone proteins, cytochrome oxidase, etc. can easily be searched by name on the internet.
- ✦ Many classical works, such as writings by Whitman, Darwin, Emerson, et al. exist in multiple print editions, and can often be accessed directly and freely online.
- ✦ Individual poems can be located online without reference to any volume.
- ✦ Included below, in alphabetical order, are author names that appear in the text.

Alberts, B. et al. *Molecular Biology of the Cell,* Garland Science. NY. 4th ed. 2002.

Alvarez, L.W., and Alvarez, W., Asaro, F., Michel, H.V. "Extraterrestrial Cause for the Cretaceous-Tertiary Extinction." Science, 1980: 4448. 1095-1108.

Anderson, Philip. "More is Different." Science, Aug. 4, 1972. V.177:4047. 393-396.

Asimov, Isaac. *Nightfall.* Original edition 1941. Included in many collections and anthologies, and available on-line.

Benton, M. "Life and Time." In *The Book of Life.* S.J. Gould, ed. W.W. Norton, NY, 1993.

Bhagavad Gita. Mascaro, Juan, Tr. Penguin Classics, 1962.

Bodhi, Bhikkhu, ed. *A Comprehensive Manual of Abhidhamma.* Buddhist Publication Society. Kandy, Sri Lanka 1993.

Bodhi, Bhikkhu. *Majjhima Nikaya.* Wisdom, Boston, 1995.

Carr, B. and Rees, M. "The Anthropic Principle and the Structure of the Physical World." *Nature,* V.278, 1979, 605-612.

Carson, Rachel. *The Sense of Wonder.* Harper and Row, 1956.

Carson Rachel. *Silent Spring.* Houghton Mifflin, 1962.

Chandrasekhar, S. *Truth and Beauty: Aesthetics and Motivations in Science.* University of Chicago Press, 1987.

Christianson, Gale E. *Edwin Hubble: Mariner of the Nebulae.* University of Chicago Press, 1995.

Darwin, Charles. *On the Origin of Species.* Many editions.

Davies, P. *Superforce.* Simon and Schuster, N.Y. 1984.

Davies, P. *The Fifth Miracle: The Search for the Origin and Meaning of Life.* Simon and Schuster, 2000.

Davies, P. *The Goldilocks Dilemma.* Houghton Mifflin, 2006.

De Duve, Christian. *Vital Dust: The Origins and Evolution of Life on Earth.* Basic Books/Harper Collins, 1995.

Delbanco, Andrew. *Melville: His World and Work.* Vintage/Random House, NY 2005.

Ehrlich, P. *The Population Bomb.* Sierra Club/Ballantine, 1968.

Einstein, A. *The Expanded Quotable Einstein.* Ed., Calaprice, A. Princeton University Press, 2000.

Einstein, A. *Ideas and Opinions.* Three Rivers Press, NY, 1954.

Eiseley, Loren. *The Firmament of Time.* Atheneum, NY, 1967.

Emerson, Ralph W. "Nature" in *Essays; First and Second Series.* Vintage/The Library of America, NY 1990, and many other editions.

Enz, Charles, P. *No Time to be Brief – A Scientific Biography of Wolfgang Pauli.* Oxford University Press, 2010.

Eraly, A. et al. *India: People, Place, Culture, History.* DK Publishing, 2008.

D'Espagnet, Bernard. *On Physics and Philosophy.* Princeton University Press, 2006.

Feynman, R. *QED: The Strange Theory of Light and Matter.* Princeton University Press, 1988.

Fraser, Gordon. *Cosmic Anger: Abdus Salam – The First Muslim Nobel Scientist.* Oxford University Press, 2008.

Frayn, Michael. *Copenhagen.* Anchor Press, 2000.

Freud, S. The *Future of an Illusion.* Doubleday Anchor, Garden City, NY 1953. (Also contained in *The Standard Edition of the Complete Psychological Works of Sigmund Freud.* London, the Hogarth Press, August 17, 1986).

Giberson, K.W. and Stephens, R.J. "The Evangelical Rejection of Reason". The New York Times, October 18, 2011.

Gilbert, Daniel. *Stumbling on Happiness.* Vintage Books/Random House, NY 2007.

Glashow, Sheldon, L. "Too Huge for the Atom, Too Tiny for the Star." In *Mind, Life and Universe.* Ed. Margulis, L. and Punset, E., Chelsea Green, VT, 2007

Gould, Steven Jay. "Preface." *The Book of Life.* W.W. Norton, NY, 1993.

Gould, S.J., and Eldredge, N. "Punctuated Equilibria: The Tempo and Mode of Evolution Reconsidered." Paleobiology, 3(2) 115-151, 1973.

Greene, B. *The Hidden Reality: Parallel Universes and the Deep Laws of the Cosmos.* Knopf 2011.

Hawking, Stephen. *Brief History of Time.* Bantam, NY, 1998.

Heschel, A.J. *I Asked for Wonder.* Crossroads, NY, 1983.

Hubble, Edwin. *The Realm of the Nebulae.* Yale University Press, 1936.

Isaacson, Walter. *Einstein: His Life and Universe.* Simon and Schuster, 2007.

James, William. *The Varieties of Religious Experience.* Many editions.

Kahneman, Daniel. *Thinking, Fast and Slow.* Farrar, Straus and Giroux, NY, 2011.

Kauffman, Stuart. *Reinventing the Sacred.* Basic Books, NY, 2008.

Kazantzakis, Nikos. *Zorba the Greek.* Simon and Schuster, 1952.

Kubler-Ross, Elizabeth. *On Death and Dying.* Macmillan, NY 1969.

Linde, A.D. "Particle Physics and Inflationary Cosmology." *Contemporary Concepts in Physics.* CRC Press, 1990.

Loewenstein, Werner R. *The Touchstone of Life*. Oxford University Press, 1999.

Mahabharata. Kishan Mohan Ganguli, Tr. Internet Sacred Text Archive.

Mahasatipattha Sutta. Vipassana Research Publication, Seattle, 1996.

Malthus, R. *An Essay on the Principle of Population*. Many editions.

Melville, Herman. *Moby Dick*. Many editions including University of California Press, Berkeley, CA 1979.

Mlodinow, Leonard. *The Drunkard's Walk*. Vintage, Random House, NY, 2009.

Monod, Jacques. *Chance and Necessity*. English Tr., Vintage, 1972.

Morowitz, H.J. *The Emergence of Everything*. Oxford University Press, 2002.

Neruda, Pablo. "Poetry." Tr. Reid, Alstair in *Neruda, Selected Poems*. Ed. Tarn, Nathaniel. Houghton Mifflin, Boston, 1970.

Newton, I. *Principia Mathematica*. Many editions.

Plato. "Phaedrus." In *The Works of Plato,* Ed. Erman, I. The Modern Library, NY, 1928. Many other editions.

Sagan, Carl. *Cosmos*. Ballantine Books, NY 1980.

Sen, Amartya. *The Argumentative Indian: Writings on Indian History, Culture and Identity*. MacMillan, 2006.

Seneca. *On the Shortness of Life*. Tr. C.D.N. Costa. Penguin, 1997.

Shubin, Neil. *Your Inner Fish*. Vintage/Random House, NY, 2008.

Steinhardt, Paul J. and Turok, N. *Endless Universe*. Doubleday/Random House, NY, 2007.

Tagore, Rabindranath. *Gitanjali*. Macmillan, London 1913. Many editions.

Tegmark, M. *The Mathematical Universe*. Cornell University Library, 2007

Thomas, Dylan. *A Child's Christmas in Wales*. Many editions.

Thomas, L. *The Lives of a Cell: Notes of a Biology Watcher*. Viking, NY, 1974.

Thoreau, Henry D. *Walden, or Life in the Woods*. Many editions.

Tudge, Colin. *The Tree*. Three Rivers Press, NY, 2005.

Wali, K.C. *Chandra: A Biography of S. Chandrasekhar.* University of Chicago Press. 1992.

Weiler, Edward J. *Hubble: A Journey Through Space and Time.* Abrams, 2010.

Weinberg, S. *The First Three Minutes: A Modern View of the Origin of the Universe.* Basic Books, 1993.

Weinberg, S. "Designer Universe?" On PhysLink.com based on a talk given at the Conference on Cosmic Design of the American Association for the Advancement of Science, April 1999.

Weisskopf, Victor. "Address of Victor F. Weisskopf." Science, March 14, 1980, p.1163.

Wells, H.G. *The Time Machine.* Many editions.

Whitman, Walt. *Leaves of Grass.* Many editions.

Wilson, E.O. *Biophilia,* Harvard University Press, 1984.

Wilson, R. "Elinor Ostrom and Camp Wopatopa." The Monkey Cage, On-line. June 13, 2012.

Wolfe, Thomas. *Look Homeward Angel.* Charles Scribner and Sons, 1929.

Woodbury, Neal. "Protein Dynamics Control the Kinetics of Initial Electron Transfer in Photosynthesis." Science. May 4, 2007.

Index

CPSIA information can be obtained at www.ICGtesting.com
Printed in the USA
BVOW08s1238141215

430229BV00001B/184/P